NJU SA 2016-2017

THE YEAR BOOK OF ARCHITECTURE PROGRAM SCHOOL OF ARCHITECTURE AND URBAN PLANNING

南京大学建筑与城市规划学院建筑系　教学年鉴

王 丹 丹 编　EDITOR: WANG DANDAN

东南大学出版社・南京　SOUTHEAST UNIVERSITY PRESS, NANJING

建筑设计及其理论
Architectural Design and Theory

张　雷　教　授	Professor ZHANG Lei
冯金龙　教　授	Professor FENG Jinlong
吉国华　教　授	Professor JI Guohua
周　凌　教　授	Professor ZHOU Ling
傅　筱　教　授	Professor FU Xiao
王　铠　副教授	Associate Professor WANG Kai
钟华颖　讲　师	Lecturer ZHONG Huaying

城市设计及其理论
Urban Design and Theory

丁沃沃　教　授	Professor DING Wowo
鲁安东　教　授	Professor LU Andong
华晓宁　副教授	Associate Professor HUA Xiaoning
胡友培　副教授	Associate Professor HU Youpei
窦平平　副教授	Associate Professor DOU Pingping
刘　铨　讲　师	Lecturer LIU Quan
尹　航　讲　师	Lecturer YIN Hang
唐　莲　讲　师	Lecturer TANG Lian
尤　伟　讲　师	Lecturer YOU Wei

建筑历史与理论及历史建筑保护
Architectural History and Theory, Protection of Historic Building

赵　辰　教　授	Professor ZHAO Chen
王骏阳　教　授	Professor WANG Junyang
胡　恒　教　授	Professor HU Heng
冷　天　副教授	Associate Professor LENG Tian
王丹丹　讲　师	Lecturer WANG Dandan
孟宪川　讲　师	Lecturer MENG Xianchuan

建筑技术科学
Building Technology Science

鲍家声　教　授	Professor BAO Jiasheng
秦孟昊　教　授	Professor QIN Menghao
吴　蔚　副教授	Associate Professor WU Wei
郜　志　副教授	Associate Professor GAO Zhi
童滋雨　副教授	Associate Professor TONG Ziyu

南京大学建筑与城市规划学院建筑系
Department of Architecture
School of Architecture and Urban Planning
Nanjing University
arch@nju.edu.cn　　http://arch.nju.edu.cn

教学纲要
EDUCATIONAL PROGRAM

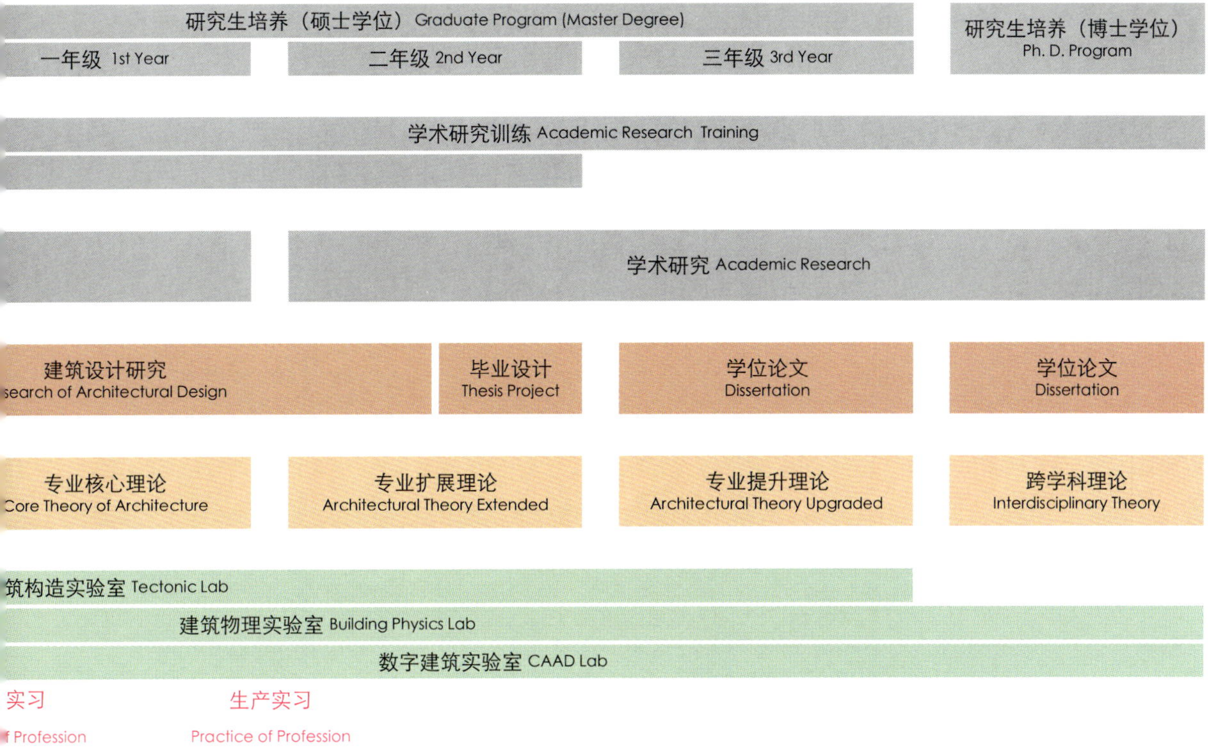

研究生培养（硕士学位）Graduate Program (Master Degree)			研究生培养（博士学位）Ph. D. Program
一年级 1st Year	二年级 2nd Year	三年级 3rd Year	

学术研究训练 Academic Research Training	

	学术研究 Academic Research

建筑设计研究 Research of Architectural Design	毕业设计 Thesis Project	学位论文 Dissertation	学位论文 Dissertation
专业核心理论 Core Theory of Architecture	专业扩展理论 Architectural Theory Extended	专业提升理论 Architectural Theory Upgraded	跨学科理论 Interdisciplinary Theory

建筑构造实验室 Tectonic Lab
建筑物理实验室 Building Physics Lab
数字建筑实验室 CAAD Lab

实习 Profession   生产实习 Practice of Profession

课程安排
CURRICULUM OUTLINE

	本科一年级 Undergraduate Program 1st Year	本科二年级 Undergraduate Program 2nd Year	本科三年级 Undergraduate Program 3rd Year
设计课程 Design Courses	设计基础 Basic Design	建筑设计基础 Basic Design of Architecture 建筑设计（一） Architectural Design 1 建筑设计（二） Architectural Design 2	建筑设计（三） Architectural Design 3 建筑设计（四） Architectural Design 4 建筑设计（五） Architectural Design 5 建筑设计（六） Architectural Design 6
专业理论 Architectural Theory	逻辑学 Logic	建筑导论 Introductory Guide to Architecture	建筑设计基础原理 Basic Theory of Architectural Design 居住建筑设计与居住区规划原理 Theory of Housing Design and Residential Planning 城市规划原理 Theory of Urban Planning
建筑技术 Architectural Technology	理论、材料与结构力学 Theoretical, Material & Structural Statics Visual BASIC程序设计 Visual BASIC Programming	CAAD理论与实践 Theory and Practice of CAAD	建筑技术（一）结构与构造 Architectural Technology 1: Structure & Construction 建筑技术（二）建筑物理 Architectural Technology 2: Building Physics 建筑技术（三）建筑设备 Architectural Technology 3: Building Equipment
历史理论 History Theory	古代汉语 Ancient Chinese	外国建筑史（古代） History of World Architecture (Ancient) 中国建筑史（古代） History of Chinese Architecture (Ancient)	外国建筑史（当代） History of World Architecture (Modern) 中国建筑史（近现代） History of Chinese Architecture (Modern)
实践课程 Practical Courses		古建筑测绘 Ancient Building Survey and Drawing	工地实习 Practice of Construction Plant
通识类课程 General Courses	数学 Mathematics 语文 Chinese 名师导学 Guide to Study by Famed Professors 计算机基础 Basic Computer Science	社会学概论 Introduction of Sociology 社会调查方法 Methods for Social Investigation	
选修课程 Elective Courses		城市道路与交通规划 Planning of Urban Road and Traffic 环境科学概论 Introduction of Environmental Science 人文科学研究方法 Research Method of the Social Science 美学原理 Theory of Aesthetics 管理学 Management 概率论与数理统计 Probability Theory and Mathematical Statistics 国学名著导读 Guide to Masterpieces of Chinese Ancient Civilization	人文地理学 Human Geography 中国城市发展建设史 History of Chinese Urban Development 欧洲近现代文明史 Modern History of European Civilization 中国哲学史 History of Chinese Philosophy 宏观经济学 Macro Economics 管理信息系统 Management Operating System 城市社会学 Urban Sociology

本科四年级 Undergraduate Program 4th Year	研究生一年级 Graduate Program 1st Year	研究生二、三年级 Graduate Program 2nd & 3rd Year
建筑设计（七） Architectural Design 7 建筑设计（八） Architectural Design 8 本科毕业设计 Graduation Project	建筑设计研究（一） Design Studio 1 建筑设计研究（二） Design Studio 2 数字建筑设计 Digital Architecture Design 联合教学设计工作坊 International Design Workshop	专业硕士毕业设计 Thesis Project
城市设计理论 Theory Urban Design	城市形态研究 Study on Urban Morphology 现代建筑设计基础理论 Preliminaries in Modern Architectural Design 现代建筑设计方法论 Methodology of Modern Architectural Design 景观都市主义理论与方法 Theory and Methodology of Landscape Urbanism	
建筑师业务基础知识 Introduction of Architects' Profession 建设工程项目管理 Management of Construction Project	材料与建造 Materials and Construction 中国建构（木构）文化研究 Studies in Chinese Wooden Tectonic Culture 计算机辅助技术 Technology of CAAD GIS基础与运用 Concepts and Application of GIS	
	建筑理论研究 Study of Architectural Theory	
生产实习（一） Practice of Profession 1	生产实习（二） Practice of Profession 2	建筑设计与实践 Architectural Design and Practice
景观规划设计及其理论 Theory of Landscape Planning and Design 东西方园林 Eastern and Western Gardens 地理信息系统概论 Introduction of GIS 欧洲哲学史 History of European Philosophy 宏观经济学 Macro Economics 政治学原理 Theory of Political Science 社会学定量研究方法 Quantitative Research Methods in Sociology	建筑史研究 Studies in Architectural History 建筑节能与可持续发展 Energy Conservation & Sustainable Architecture 建筑体系整合 Advanced Building System Integration 规划理论与实践 Theory and Practice of Urban Planning 景观规划进展 Development of Landscape Planning	

1—157 年度改进课程 WHAT'S NEW

2
设计基础（一）
BASIC DESIGN 1

46
建筑设计（五+六）：城市建筑：社区中心
ARCHITECTURAL DESIGN 5 & 6: URBAN ARCHITECTURE: COMMUNITY CENTER

12
设计基础（二）
BASIC DESIGN 2

58
本科毕业设计：长汀历史名城更新与建筑设计
GRADUATION PROJECT:RENEWAL AND ARCHITECTURAL DESIGN OF CHANGTING HISTORICAL CITY

22
建筑设计（一）：老城住宅设计
ARCHITECTURAL DESIGN 1:RESIDENTIAL DESIGN OF OLD TOWN

76
本科毕业设计：数字化设计与建造
GRADUATION PROJECT:DIGITAL DESIGN & BUILDING

34
建筑设计（三）：赛珍珠纪念馆扩建
ARCHITECTURAL DESIGN 3:EXPANSION OF PEARL BUCK MEMORIAL

88
建筑设计研究（一）：记忆 · 场所 · 叙事
DESIGN STUDIO 1: MEMORY · PLACE · NARRATIVE

96
建筑设计研究（二）：建构研究"低技建造"设计研究
DESIGN STUDIO 2: DESIGN RESEARCH ON "LOW-TECH CONSTRUCTION" IN CONSTRUCTION RESEARCH

108
研究生国际教学交流计划
THE INTERNATIONAL POSTGRADUATE TEACHING PROGRAM

131—177　附录 APPENDIX

131—143　建筑设计课程 ARCHITECTURAL DESIGN COURSES

145—147　建筑理论课程 ARCHITECTURAL THEORY COURSES

149—151　城市理论课程 URBAN THEORY COURSES

153—155　历史理论课程 HISTORY THEORY COURSES

157—159　建筑技术课程 ARCHITECTURAL TECHNOLOGY COURSES

161—169　回声——来自毕业生的实践 ECHO—FROM PRACTICES OF GRADUATES

171—177　其他 MISCELLANEA

年度改进课程
WHAT'S NEW

设计基础（一）
BASIC DESIGN 1
季鹏

  设计基础作为设计相关学科的启蒙课程，一直以来都是引导学生进行学科认知与思维模式转换的重要教学环节，同样也是建筑学基础课程的重要组成部分。在教学实践中，如何合理安排通识基础与专业基础知识的比重、如何在确保课程体系科学性的基础上融入趣味性与实验性、如何将艺术与科技有效结合来制定教学内容，在多元文化的当下，对于建筑学科的宽口径人才培养模式，具有积极的影响与深远的意义。

  近年来，与南京艺术学院设计学院的设计基础课联动教学，是充分体现学科优势互补、强强联合的有益尝试。在双方教学团队的共同努力下，经过不断的教学探索与改革后，课程体系不断完善并逐步成型。课题从艺术设计过渡到建筑设计基础，从平面到立体，从独立作业到团队协作。课程总体由7个循序渐进的课题组成，训练内容涉及综合绘画、材料拼贴、尺规作图、陶瓷烧造、丝网印刷、观念摄影、模型制作、图表分析、立体构成、空间建构与设计软件学习。通过两个学期的课程安排，为建筑学专业的一年级新生开启了一个全新的学科领域，为今后的专业学习夯实了基础。

  本课题是系列课题的第一部分：看到世界的美。

  作为造型训练的基础课程，课题1主要培养学生理解同一形式的多种表达方式。自行车是学生们熟悉的日常用品，但是复杂的点线面结构所体现的机械美学，在艺术表现时又充满了挑战。本课题以自行车为研究对象，通过素描、拼贴、色彩、陶瓷烧造与丝网印刷等多种艺术表现手法，拓展学生观看的视角以及造型、色彩与材料的基本表达方式。

As the enlightenment course of design-related subject, Basic Design is always important teaching link to guide students for subject cognition and thinking mo conversion, and also the important part of architecture foundation. In the teach practice, how to rationally arrange the proportion of general knowledge basis a professional basic knowledge, how to merge interest and experiment on the ba of ensuring scientific course system, how to effectively combine art with technolo to confirm teaching content, under the background of diversified culture, it h positive influence and profound meaning to the wide-caliber talent cultivation mode architecture subject.

In recent years, the teaching combined with the design basis course of School Design of Nanjing University of the Arts is a win-win beneficial try to fully refl complementary subject advantage. Under the common efforts of teaching teams both sides, through constant teaching exploration and reform, the curriculum syste is constantly improved and took shape gradually formed. The tasks are transit from art design to architectural design basis, from plane to solid, from independe operation to teamwork. The overall curriculum consists of 7 progressive tasks, a the training content involves comprehensive drawing, material collaging, drawi with ruler and compass, ceramic firing, silk-screen printing, conceptual photograp modeling, chart analysis, three dimensional composition, spatial construction a design software learning. Two semesters of course arrangement open a bra new subject field for the freshmen, and lay a solid foundation for the subseque professional learning.

This subject is the first part of series of subjects: see the beauty of world.

As the basic course of modelling training, task 1 mainly cultivates students understand various expression methods of the same form. The bicycle is an arti for daily use familiar to students, but the mechanical aesthetics reflected from t complex point-line-plane structure is full of challenge in artistic expression. In t task, with the bicycle as the research object, through various artistic expressi methods including sketch, collage, color, ceramic firing and silk-screen printing, et view of students and basic expression methods for modeling, color and material a developed.

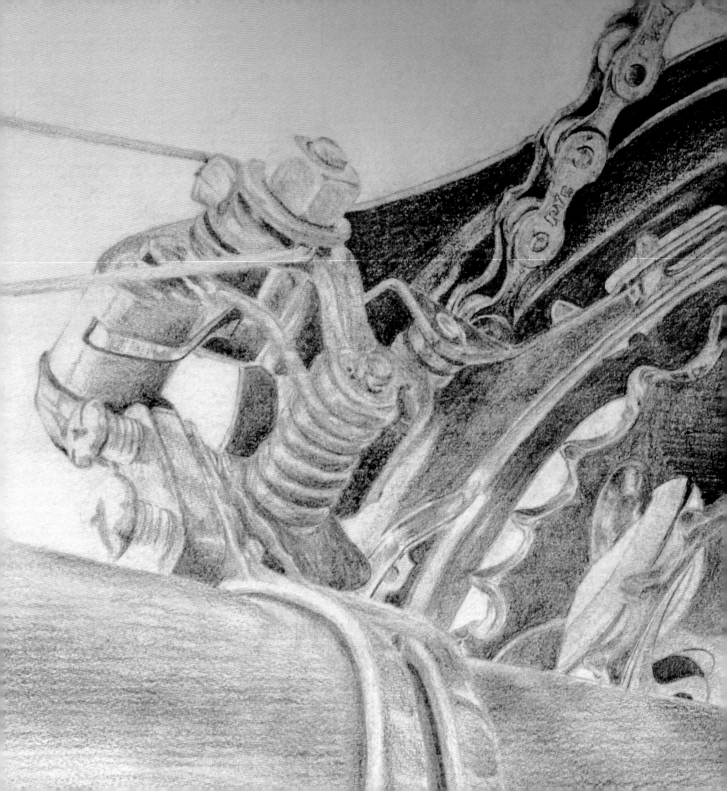

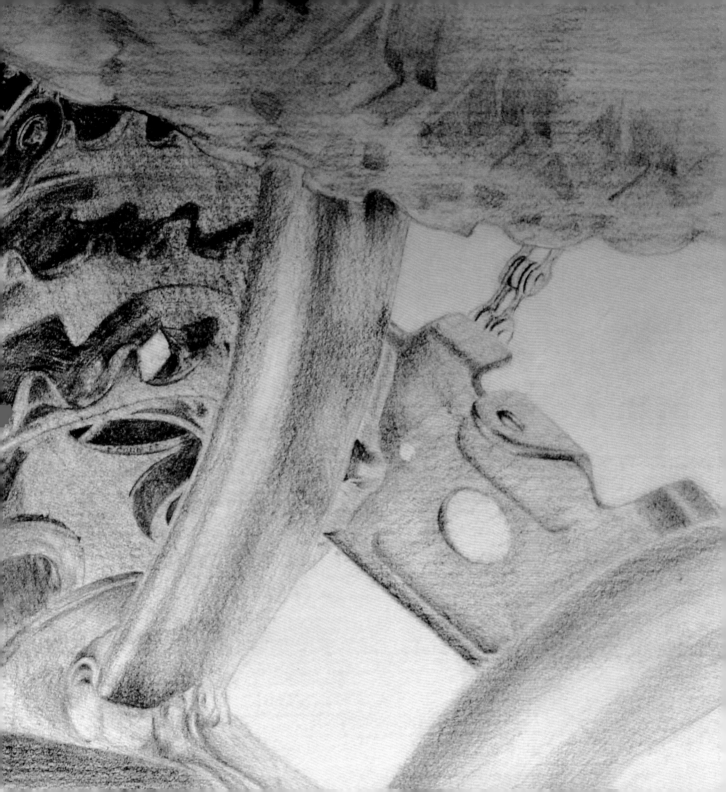

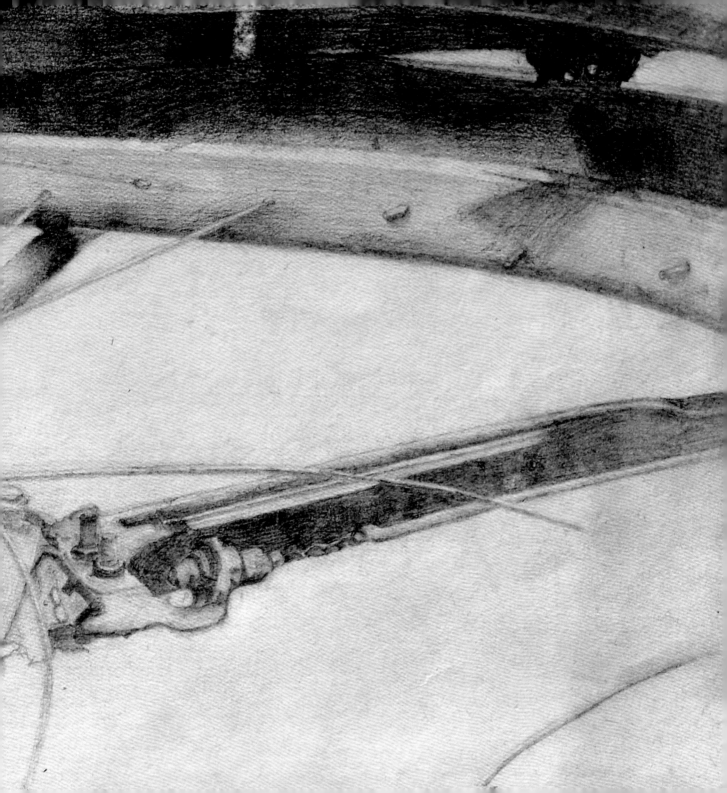

设计基础（二）
BASIC DESIGN 2

丁沃沃 鲁安东 唐莲 刘妍

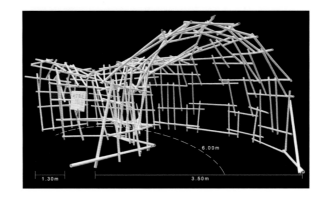

作为本科生一年级第二学期空间训练的第三部分，即最后一个练习，在"动作—空间分析"以及"折纸—空间包裹"之后，作为对材料与造型思维能力的进一步训练，练习三"空间搭建——互承的艺术"将设计训练的空间人体局部的空间运动扩大到人体行动路径的尺度，通过真实搭建身体能够进入或通过的空间结构，训练学生建立对建筑的材料结构与建造施工的初步认识。

1.课程设置

"空间搭建——互承的艺术"教学历时六周，要求以指定的结构语言（互承结构）、指定材料的杆件（4~5cm直径的PVC管），以穿孔绑扎为节点形式，在指定的场地，以有限的材料数量（100根左右杆件），完成一件特定高度与开口尺度的结构装置的设计与建造。课程在"基于场地的形式操作"大主旨下，将结构的尺度扩大到真实的人体行进运动空间。在近于建筑尺度的建造中，结构面对材料的自重、强度以及外部的自然与人为荷载的环境，面对真实的建造施工问题以及结构失效的压力。为此，教学过程设置三个阶段的练习。第一阶段是基于模型杆件的基础练习（一周），第二阶段是以1:10的模型杆件进行结构形式的设计（三周），第三阶段则使用足尺材料进行搭建（二周）。

1.1 互承结构基础练习

阶段一互承结构基础练习向学生传授互承与非互承结构的基本原理，训练学生对于材料与结构特性的认知。学生需要认识与学习互承结构的基本单元形式、构造特征与设计参数，并且使用木质模型材料，对选定的单一形式的结构单元加以扩展，制作一个半径15~20cm的匀质穹体。

这个练习有助于使学生迅速建立起对于互承结构——这种特殊的结构形式——的构形特征与结构原理的基本认识。当调整杆件与单元格网的尺度参数，结构体的角度（曲率）产生相应的、难于准确计算的空间几何变化。学生被要求对杆件与单元尺度进行调整尝试，并对杆件与结构的几何关系进行推算、测量记录与总结归纳，定性掌握这种结构的设计参数与几何形式之间的关系。这一训练延续了对于材料认知的训练。学生将通过模型制作，切实体会具有较高刚性的短小杆件材料与具有一定弹性变形能力的稳定结构体之间的转变。

1.2 互承单元的变形与组合研究

在阶段一建立起来的对于互承结构单元特征与参数调控的认识基础上，在阶段二与阶段三，学生将面对特定的任务要求进行互承结构的造型设计。设计任务为建造一个非对称形式的互承结构曲面覆盖体，建造高度不小于2m，并带有使人可以正步行进的通行孔洞。其中阶段二使用木质模型杆件进行1:10的模型设计，阶段三使用实际建造材料（PVC杆件）进行现场建造。

在阶段二，学生以前一阶段建立的对于单一单元造型性质的认识为基础，对同形式、不同尺度的互承单元进行组合与变形，以此为造型手段。对造型提出设计概念后，利用分析图示与模型进行造型探讨与结构推敲。在这一阶段，学生极大地面形式概念与结构实现之间的挑战。受到互承结构的曲线造型的鼓舞，学生会设想灵而生动的几何意向。但在模型结构面前，则要面对几何构造的限制与结构稳定性的力。这一阶段三周的教学首先以两人的小组进行设计，之后相似的设计被合并，失的设计被淘汰。在教师的指导下，学生会在动手的过程中不断进行调整改进模型，终形成5组设计主题，每组4~6名成员。

除了通过模型来呈现变形与组合的可能之外，还需手绘图解单元几何尺寸、系。对空间网格的基础原型以及变形进行图示解析，了解单元与节点的变形与组合正反方向互承等因素对于形式控制的意义。绘图训练有两个层面的目标，一方面，构设计作为理性思考与控制的产物，要求学生从平面投影图示入手进行概念与初步式思考。另一方面，建造成果要求具有"可复制性"，即模型设计需要得到准确的像记录与图纸表达。

1.3 空间搭建：互承结构的实体建造

因为阶段二的模型设计成果即直接应用于阶段三的实际建造，材料的直径、度、数量、节点（打孔）位置，必须得到准确的统计与记录。尤其本年度的搭建教中重复使用前面两年遗留的搭建材料，学生面对更加严苛的材料条件，即相应的成计算训练。

在阶段三，在两周的时间中，学生需将1:10模型"放样"到足尺的PVC管材材上，使用打孔绑扎节点进行固定。在这个过程中，学生将对真实建造与结构体产生

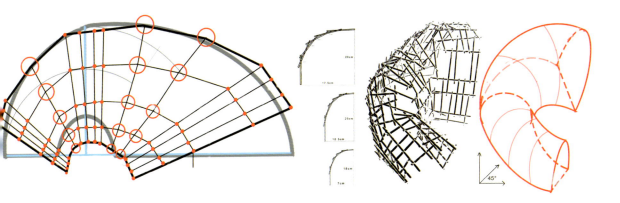

的直观体验：在材料层面，PVC管材较之模型木材更加柔软，穿孔绑扎节点的转动力更大，材料的自重效应更加显著，因此被小尺度模型掩盖的结构问题会被突显，缺陷的结构会面对更加苛刻的考验。在构造层面，虽然可以被绑扎的灵活性掩盖，节点构造精度的误差被远远放大强调。在施工层面，施工步骤与组织直接决定了结构的可操作性，团队分工合作能力也得到训练。在结构稳定性层面，学生在第一周在场地完成搭建后，成品在答辩展示前展陈于场地（平台广场）。在这一周内，要接受种种外力——风、雨与好奇民众的荷载考验。稳定性有缺陷的结构，在此期间会产生大变形，学生不得不在几天后返回现场，调整并加强。在答辩之后，变形最大的小品被要求进行结构失效分析。在最终的答辩与评判中，模型设计的应用、互承结构特点、结构的复杂度与难度、成品对设计概念的实现、成品结构的稳定性等，都是评判作品是否优秀的考虑因素。

2．教学成果与讨论

"空间搭建—互承的艺术"课程取得了较好的教学效果，学生兴趣浓厚，最终作品及图纸完成度较高。对于一学期的空间基础教学来说，"空间搭建"承接了"动—空间分析"中学生对身体尺度及空间关系的认知，以及"空间包裹"中场地—形与结构的关系，并将空间与造型的训练进一步推向真实建造，真切、现实而又精简地将一年级学生引入"建造"这一建筑学之根本核心问题。

在本课程的最后的高潮——"艺术与理性"成果展中，5组装置根据特定的路线设布置在场地，学生身着练习二的"折纸服装"，以设计的路线、动作，在特别选定背景音乐下，通过构造的洞口穿过5组装置：在亲手建造的"舞台布景"中，材料、构、节奏、空间融为一体，融合体现三部分课程的整体训练。

在空间搭建的训练中，以简练甚至抽象的方式，涉及现实建造的一些核心问题：地、材料、构造、结构、形式、理性与偶然性、成本、不利荷载、变形、失效与加以互承结构这一相对难于计算控制的非传统结构作为训练介质，材料、形式与结之间的关系得到了极为突出的强调。

As the third part of space training of freshmen in the second semester, that is, the last exercise, after "motion-space analysis" and "paper folding-space wrapping", as the further training on material and modeling thinking ability, exercise III "space building—art of reciprocal structures" will expand space human body´s local space motion of design training to human body motion path, through really building the space structure that body can enter or pass, training students to establish preliminary cognition of building material structure and building construction.

1. Curriculum

"Space building—art of reciprocal structures" lasts for six weeks, requires students to complete the design and building of structure device with specific height and opening size by limited material (about 100 pieces of rod) on the designated site in the node form of perforated binding in designated structure language(mutual-bearing structure), rod of designated material (4~5cm PVC pipe). Under the purport of "form operation based on site", the course expands structure dimension to real human body motion space. In the building close to the architecture dimension, the structure faces deadweight, strength of material and external natural and manmade load environment, real construction problem and pressure of ineffective structure. Therefore, three stages of exercise are set in the teaching. The first stage is basic exercise based on model rod (one week); the second stage is design of structure form of 1:10 model rod (three weeks); the third stage is building by full-size material (two weeks).

1.1 Basic exercise of reciprocal structures

First stage, basic exercise of reciprocal structures, teaches students the basic principle of reciprocal and non-reciprocal structures, and trains the cognition of students on material and structure characteristics. Students are required to know and learn the basic unit form, structure characteristics and design parameter of reciprocal structures, use wooden model material to extend selected structure unit in a single form, and make a homogenous dome with the diameter of 15~20cm.

This exercise can help students rapidly establish the basic cognition of forming characteristics and structure principle of the special structure form—reciprocal structures. When the dimension parameter of rod and unit grid is adjusted, the angle (curvature) of structure will have relevant space geometric change hard to precisely calculate. Students are required to try to adjust the dimension of rod and unit, calculate geometric relation of rod and structure, measure record and summarize, qualitatively master the relationship between design parameter and geometric form of such structure. This exercise continues the training on material cognition. Students will personally feel the change between short and small rod material with high rigidness and stable structure with certain elastic deformation capacity through modeling.

1.2 Deformation and combination research of reciprocal unit

On the cognition basis of unit characteristics of reciprocal structures and parameter adjustment and control established at the first stage, at the second stage and the third stage, students will face specific task requirement for modeling design of reciprocal structures. The design task is building an asymmetrical cambered covered body in reciprocal structures, with the building height not less than 2m, and a hole that human can walk in. Wooden model rod is used for 1:10 model design at the second stage, actual building material (PVC rod) is used for site building at the third stage.

At the second stage, students base on the cognition of single unit model nature established at the previous stage, combine and deform the reciprocal units in different forms and dimensions, as the modeling method. After design concept is proposed to the modeling, analysis cutline and model are utilized for building discussion and structure deliberation. At this stage, students face great challenge between form concept and structure realization. Inspired by the curve model of reciprocal structures, students will conceive flexible and vivid geometric intention. However, in front of the model structure, students shall face the limit of geometric structure and pressure from structure stability. In three-week teaching at this stage, two students in a group create a design, then the similar designs will be merged, the failed design will be washed out. Under the guidance of teaches, students will constantly adjust and improve model in the practice. Finally 5 groups of deign theme will be formed, with 4~6 members in each group.

Besides showing the possibility of deformation and combination through model, students shall also draw to solve the geometric dimension and relation of unit. Students shall use graphs to analyze the basic prototype and deformation of space grid, understand the meaning of deformation and combination of unit and node, reciprocal structures in forward and backward directions, etc. to form control. Drawing training has two objectives, on one hand, structure design is the product of rational thinking and control, that requires students to start with plane projection cutline for concept and preliminary thinking. On the other hand, the building result shall be "reproductive", that is, the model design shall have precise image record and drawing expression.

1.3 Space building: solid building of reciprocal structures

The model design result at the second stage is directly applied in the actual building at the third stage. Material diameter, length, quantity, node (hole) position must have precise statistics and record. Especially in the building teaching in this ye the building material left over from the past two years is used. Students face mo rigorous material condition, that is, relevant cost calculation training.

At the third stage, in two weeks, students shall "set out" 1:10 model on full size PVC pipe material, use punched bound node for fixing. In this process, students v have preliminary experience in real building and structure: on the level of materi PVC pipe material is softer than wood, the rotation capacity of punched bound no is larger, the deadweight effect of material is more obvious, therefore, the structu problem of being covered by small size model will be highlighted, defective structu will face more rigorous test. On the level of structure, although being covered the flexibility of binding, the precision error of node structure is far emphasized. C the level of construction, construction step and organization directly decides t operability of structure, team division and cooperation capacity is trained. On the lev of structure stability, after the building is completed on the site in the first week, t finished product of students is displayed on the site (platform square) before defens In this week, the finished product will receive the test from external force—win rain and curious public. Structure with defective stability will have large deformatio during this period, students have to return to the site to adjust and reinforce a fe days later. After defense, the group with the largest deformation will be required analyze structure failure. In the final defense and judgment, application of mod design, characteristics of reciprocal structures, complexity and difficulty of structur realization of design concept on the finished product, stability of finished structur etc. will be the considerations for judging an excellent work.

2. **Teaching result and discussion**

"Space building—art of reciprocal structures" achieves a good teaching effec students have strong interest, and the completion degree of final work and drawir is high. To the space basic teaching in the first semester, "space building" carries the cognition of students on body dimension and space relation in "motion-spac analysis", and relationship of site-form and structure in "space wrapping", furth promotes the training of space and modeling to real building, really and simpl introduces freshmen to the core issue of "building" in the architecture.

In the last climax of this course—"art and rationality" achievement show, 5 groups devices are laid on the site according to specific route, students wear "paper folde clothing" in exercise II, with designated route, motion, under the specific backgroun music, pass 5 groups of devices through the opening of structure: in the personal built "stage setting", material, structure, rhythm, space are merged to reflect th overall training in the three parts of course.

In the training of space building, in the simple and even abstractive form, som core issues of real building are involved: site, material, structure, form, rationali and contingency, cost, unfavorable load, deformation, failure and reinforcemen Reciprocal structures, this non-traditional structure hard to calculate and control, used as training medium to extremely emphasize the relationship among materi form and structure.

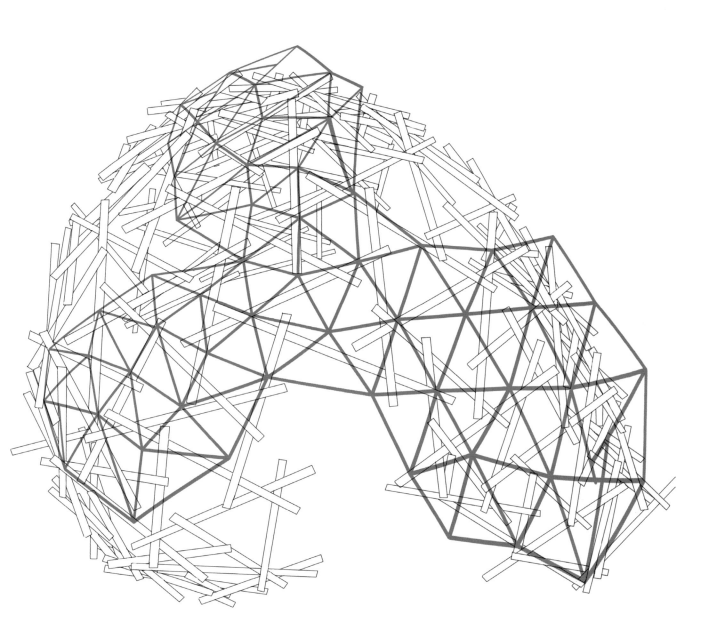

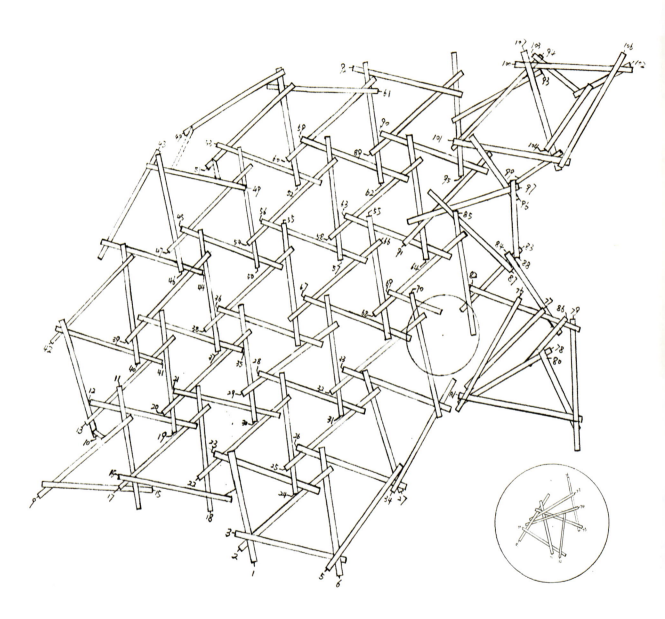

空间搭建的训练中，以简炼甚至抽象的方式，涉及现实建造的一些核心问题：场地、材料、构造、结构、形式、理性与偶然性、成本、不利荷载、变形、失效与加固。

In the training of space building, in the simple and even abstractive form, some core issues of real building are involved: site, material, structure, form, rationality and contingency, cost, unfavorable load, deformation, failure and reinforcement.

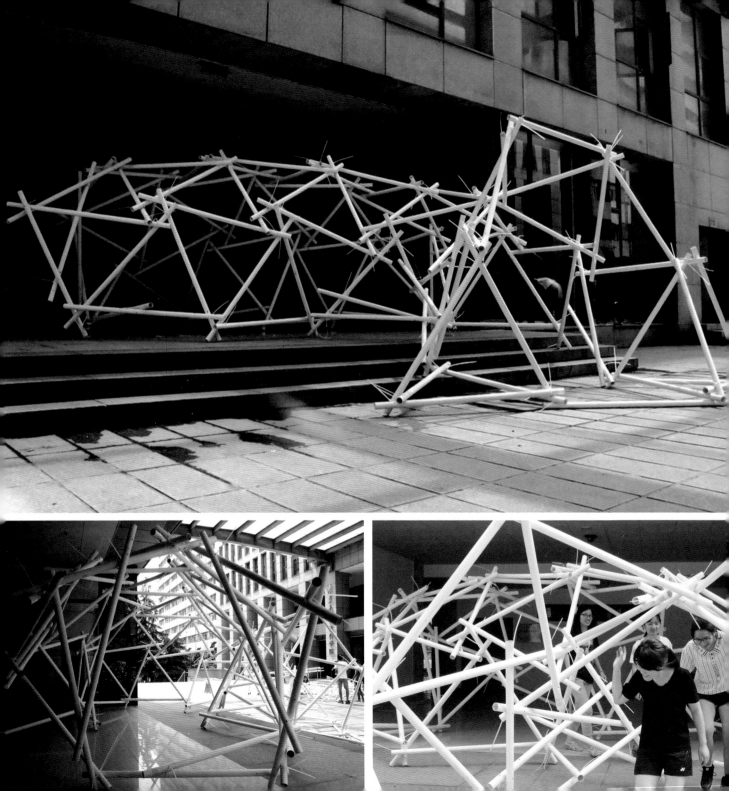

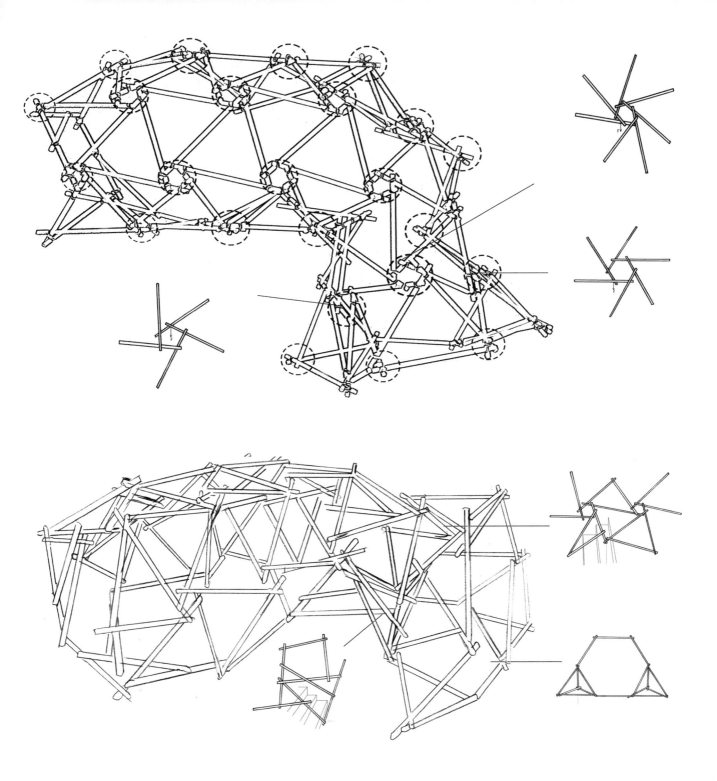

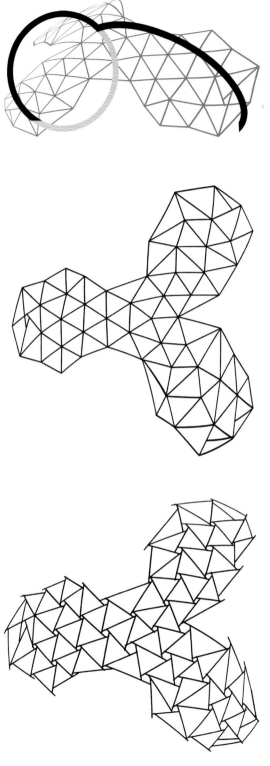

建筑设计（一）ARCHITECTURAL DESIGN 1
老城住宅设计
RESIDENTIAL DESIGN OF OLD TOWN

冷天 刘铨 王丹丹

南京大学建筑与城市规划学院的建筑学教育，在"拓宽基础、分流培养"的基本方针指导下，确立了以两年本科通识教育为基础、两年专业教育为主干、两年专业硕士教育为出口的特色人才培养模式。而这种模式也要求在教案设置和课程展开上，应结合国际建筑理论与学术的发展趋势，并紧密配合中国社会发展产生的各种建筑和城市问题。因此，为更好地结合现实建筑设计问题，建筑设计（一）"老城住宅设计"的教案，尝试将尺度与感知的空间操作引入本科设计课程体系之中，从课程背景、教学重点、典例分析、设计操作、经验教训等方面展开，探索建筑设计基础教学的新思路。

1.课程背景

工业革命以来，随着工程技术的进步、建造材料的拓展、建筑类型的多样以及建筑功能的复杂化与复合化，从包豪斯到德州骑警，从库柏联盟到ETH，建筑学专业教育的内容已经被极大地拓展，不再局限于"学院派"那种热衷于形式美的训练和单一的表达模式，更多地关注社会需求的多样性与复杂性、工程技术和材料应用的合理性与经济性等方面。

建筑设计初步是在设计基础的课程上构建的，经过了前期必要的、基本的知识储备阶段，开始真正学习具体的建筑设计，是整个建筑设计课程教学体系的入门阶段和启蒙阶段。设计初步的特点主要表现在：初始性、基础性和被动性。学生在建筑设计学习的开始环节，尚未形成独立的判断和评价能力，因此教案的设置必须着眼于设计内容、步骤、方法和表达基本概念的建立，结合每个特定设计阶段，强调针对性的训练重点，从建筑形式表达、绘图训练走向以问题为导向的分阶段设计训练。此外，建筑设计教学体系还必须处理好课程设置的系统性、关联性和逻辑性，不断调整基础课程、专业基础课程和专业课程之间的关系。

建筑设计（一）老城住宅设计课程的教学目的，是让学生综合运用建筑设计基础课程中掌握的建筑知识和表达工具，初步体验一个完整的小型建筑方案设计过程。训练的重点在于内部空间的整合性设计，同时希望学生在设计学习开始之初，能够主动去关注场地与界面、空间与功能、流线与出入口、尺度与感知等设计要素之间的紧密关系。

2.教学重点

建筑设计的操作过程，应该围绕不同层面的建筑问题展开，并寻找解决建筑问题的最佳整合方案。建筑设计课程教案的设置，也逐渐破除了原先以建筑类型为导向的设定，转变成当今以建筑问题为导向的设定。本课程作为建筑学本科生独立完成的一个设计任务，在尚不具备相关建筑结构、构造、技术等配套专业知识的前提下，设计训练的重点设置是基于尺度和感知的空间操作，以期在设计教学的起步阶段，引导学生理解与体会现代建筑空间的概念和内涵。

尺度：以人为尺度的空间需要研究个人和他所使用的空间、物体之间的物理关系，甚至心理关系，目标是根据人的姿势和动作来改善使用的情况。建筑和人体会在人体工学的层次上实际相遇，"人的尺度"是个相对模糊但十足意识形态的词汇，大小、形式和姿势之间的关系，才是它的本质所在。尺度首先指的是营造物与人体的相对度量关系，如何使用尺度，能否在身体上（走进去、在周边移动、待在里面）或心理上（接受它成为一种符号、一个场所或一种再现）支配它与控制它。此外，出于美学和象征上的理由，常常将人体结构和建筑进行类比，人们对于宇宙秩序的主观再现，以人体而非原子作为主要的参考点，以身体为基础来比较大与小、几何与无几何、硬与软、宽与窄、强与弱。人类对美的感知从来都与人体形式密不可分。

感知：对于拥有五感可自由支配的个人而言，建筑的体验首先是视觉和动觉的，相对次要的是听觉、嗅觉和触觉，但建筑物如果无法满足其中任何一项，都会大大损该作品的品质。人类的五感不仅是单纯的生理机能，也是可以学习的技巧，人的智力、学习和记忆能力，会把这些感官运作与自身特殊的经验、文化和时代联系起来。身体的移动确实能够帮助人类衡量物体和空间的大小。穿行、造访、舞动、抒达……这些动作都能让人类对大小尺度有更好的领略，并探索一些隐而不见的部分。趋近、离开、绕转、爬上爬下、进进出出……这些动作全都邀请人们进入一个既定的环境，自行将想要看的、听的、闻的、尝的和碰的东西组织起来。在素描和照片里，建筑无非就是影像，然而一旦盖好，它就变成各种体验和移动的场景，甚至是一连串感受组成的情节。

空间操作：哲学家和数学家把空间界定成由物品组成的星座，这样的空间往往

22

于人类所处的地面之外，落在意识经验和抽象中间。空间必然是中空的，由外部给定边界，在内部填充事物，而建筑空间是被物质界定的非物质部分，把世界的某一部分划分出来，让它适合人类的居住和使用，这正是建筑设计的本质所在。现代建筑学的一大突破，就是把建筑的空间彰显出来，进而理解空间配置的来龙去脉。对建筑师而言，介于地面、墙面和天花板之间的空缺可以是任何东西，但绝非虚无；事实上，正是建筑师采取行动的基础，把这个中空创造成容器，给它一个明确的形式，为场所提供形貌以及人们所需的移动自由。现代主义出现之后，建筑的空间在操作上具有了新的意义。有机性：局部从属于整体的含义；流动性：同以往封闭或开敞不同——流动的、贯通的、隔而不离的空间；隐喻性：对建筑空间进行塑造达到象征性的效果。

3. 典例辅助

本课程选取了两个经典的小型独立式住宅作为案例，引导学生通过大比例实体模型（1:50）的制作、分析性图纸的绘制，在空间、结构、流线、尺度等方面进行认知学习。两个案例均位于日本东京的城市中心区，在场地和功能定位上同本课程的设定均有较好的契合和参考价值。

(1) 团子坂住宅：由妹岛和世设计，占地面积24.3m²，建筑面积50.4m²。木框架结构体系，让家庭生活自身来勾勒房间的轮廓，通过标高的细微差别来区别各个空间，取消传统的隔墙界定，增添了内部空间的丰富性和趣味性。

(2) 土桥住宅：由妹岛和世设计，占地面积72m²，建筑面积60m²。钢框架结构体系，错动的楼板开口营造出一通高的中庭空间，为室内引入充足的光线，开放的概念在视觉上联系了每个楼层，由相互平行的楼梯进入，人们在空间中移动时会感到结构的变化。

上述两个典例，暗含了现代空间组织的三个策略：结构性空间、空间体量设计和自由平面。结构性空间指承载结构的秩序与其生成的空间图形之间，具有严谨一致的关系，这种研究的一致性创造出不具暧昧性的清晰构图。空间体量设计是阿道夫·路斯和约瑟夫·弗兰克在20世纪初创造出的名词，是一种构内部空间的方式，根据场所的性质和目的决定空间的高度，结构成为外部的支撑力量、为内部氛围服务的一套框架。自由平面既非作品的无政府状态也没有否定秩序，这种空间组织技术在20世纪上半叶发展出来，特别强化空间之间的相互渗透，而不是房间的并置、对齐或在剖面上堆叠。

4. 设计内容

课程设计的场地选择在学生前期进行过环境认知调研的城市历史街区中，抽出3个具有代表性的地块，并设定了相应的用地红线范围，学生选择其一完成设计。单个地块的面积在60~70m²左右，单面或相邻两面临街，周边为1~2层的传统民居。建筑功能为小型家庭独立式住宅，家庭主要成员（一对年轻夫妇）均具有从事建筑相关行业的职业背景，育有两名未成年子女。新建建筑面积不超过100m²，建筑高度≤8m（指计入容积率的空间内，不包括女儿墙，不设地下空间），建筑应包括起居室、餐厅、主卧室、儿童房、多功能房（含工作空间和客床）、厨房、餐厅、卫生间（1或2处）以及必要的储藏空间。另外，新建筑出入口附近应考虑至少两个以上电动自行车的停车及充电空间。

在教学进度上，本次课程共分为两个阶段。第一阶段是典例分析（3周），通过大比例实体模型，在空间、结构、流线、尺度等方面对两个经典住宅案例进行体验和认知。第二阶段是设计操作（5周），通过专业的平立剖图纸，配合不同比例的实体模型以及剖透视、分析图、效果图等，完成整个课程设计任务。

5. 平行思考

除了教案设置的教学重点之外，还要求学生在下列四个方面进行平行的思考和研究，最终以一个整合性的设计方案予以体现。

场地与界面：建筑的形体是其内部空间的反映，而建筑的外部空间指建筑周围或建筑物之间的环境。场地从外部限定了建筑空间的生成条件，需要结合基地现状条件对场地内的建筑、道路、绿化等构成要素进行全面合理的布置，通过设计使场地中的建筑物与其他要素形成一个有机整体发挥效用。而建筑形体的外立面，作为限定城市外部空间的垂直界面，直观地反映着城市中建筑之间的关系，舒适的城市公共开敞空间也首先依赖于合理的建筑形体与布局。

空间与功能：建筑设计最根本的目的，是获取合乎使用的空间。现代建筑日趋复杂的功能要求、建造技术和材料的突破，为建筑师创造建筑空间提供了更多可能，空间意义也成为现代建筑最重要的内涵，其意义远大于建筑内部可供使用的房间。现代建筑设计很大程度上是建筑空间的设计，建筑形式语言和设计方法均以空间作为主题展开。而建筑的功能指建筑物内外部空间应满足的实际使用要求，回答了建筑基本使用目的的问题。

流线与出入口：建筑的流线俗称动线，是指人流与车流在建筑中活动的路线，根据不同的行为方式把各种空间组织起来。一方面，建筑内部各功能空间需要合理的水平、垂直交通来相互沟通与联系；另一方面，建筑的内部空间需要考虑与场地周边环境条件的合理衔接，如街道界面的连续性、出入口位置的选择与退让处理、周边建筑外墙界面（包括其上的外窗）对新建建筑的影响、建筑之间的间距与视线干扰、日照的合理使用等。

尺度与感知：建筑内部的空间是供人来使用的，因此建筑中的各功能空间的尺度，都必须以人体作为基本的参照和考量，并结合人体的各种行为活动方式，来确定合理的建筑空间尺寸。在空间形式处理中注意通过图示表达理解空间构成要素与人的空间体验之间的关系，主要包括尺度感和围合感。建筑尺度还受到建造条件的限制，并与环境存在参照的关系。

6. 时间节点

第一周：理解典例图纸资料，制作工作模型。调研场地，分组制作场地模型（底座60cmx60cmx5cm），用1:50实体工作模型构思初步方案。

第二周：制作典例结构模型，绘制典例剖透视分析图。深化场地调研，深化初步方案。

第三周：完成典例正式模型。用1:50手绘平立剖图纸，研究功能与空间、流线与尺度，通过照片拼贴、沿街透视研究场地和界面的关系与效果。

第四周：确定基本设计方案，推进剖、立面设计，利用工作模型辅助。

第五周：深化设计方案，细化推敲各设计细节，并建模研究内部空间效果。

第六周：深化1:50图纸，开始照片拼贴效果图的制作（集中挂图点评）。

第七周：制作1:20剖透视和各分析图，制作1:20大比例模型（集中挂图点评）。

第八周：整理图纸、排版并完成课程答辩。

7. 成果要求

A1灰度图纸2张，纸质表现模型1个（比例1:20），工作模型若干。图纸包括：

(1) 总平面图（1:200），各层平面图、纵横剖面图和主要立面图（1:50），内部空间组织剖透视图1张（1:20）。

(2) 设计说明和主要技术经济指标（用地面积、建筑面积、容积率、建筑密度）。

(3) 表达设计意图和设计过程的分析图（功能分析、流线分析、结构体系等）。

(4) 纸质模型照片与电脑效果图、照片拼贴等。

Architecture education of School of Architecture and Urban Planning of Nanjing University, under the basic guideline of "broadening foundation, shunting the cultivation", establishes characteristic talent cultivation mode with two-year undergraduate general education as the foundation, two-year professional education as the backbone, two-year professional master education as the outlet. This mode also requires combining international architecture theory with academic development tendency on the basis of teaching plan setting and course, closely cooperating with various architecture and urban problems produced from Chinese social development. Therefore, to better combine with real architectural design problem, the teaching plan of Architectural Design I "Residential Design of Old Town" attempts to introduce space operation of dimension and perception into the design curriculum system, focus on course background, teaching essential, typical case analysis, design operation, experience and lesson, etc., and explore new idea of architectural design basic teaching.

1. Course background

Since the industrial revolution, with the progress of engineering technology, extension of building material, diversification of building type, complex and compound building function, from Bauhaus to Texas Rangers, from Cooper Union to ETH, the content of architecture education has been greatly developed without being limited to the training of formal beauty and single expression mode that "academism" is enthusiastic about, it concerns more about diversified and complex social demands, rationality and economy of engineering technology and material application.

Architectural design is preliminarily constructed on the course of design basis. Through early necessary, basic knowledge reserve, the real specific architectural design is started, as the elementary stage and enlightenment stage of the whole architectural design teaching system. The preliminary characteristics of design include: initial, basic and passive. At the beginning of architectural design learning, students do not form independent judging and evaluating ability, therefore, the setting of teaching plan must focus on establishing the basic concept of design content, step, method and expression, combine with each specific design stage, emphasize targeted training focus, go from expression of architectural form, drawing training to problem oriented stage design training. In addition, the teaching system of architectural design must also well deal with the system, correlation and logic of curriculum, constantly adjust the relationship between basic course, professional basic course and professional course.

The teaching purpose of Architectural Design I "Residential Design of Old Town" is to make students comprehensively apply building knowledge and expression tool in the basic course of architectural design, preliminarily experience the design process of a complete small architectural plan. The training focuses on the integration design of internal space, while students are expected to actively concern the close relation between site and interface, space and function, flow line and access, dimension and perception, etc. at the beginning of design learning.

2. Teaching essential

Operation process of architectural design shall focus on building problems of different levels, and search the optimal integration plan to solve the building problem. The setting of teaching plan of architectural design also gradually breaks the original setting oriented by building type and turns to the setting oriented by building problem now. As the first design task independently completed by architectural undergraduates, on the premise of not having supporting professional knowledge of building structure, construction, technology, etc., the focus of design training in the course is set as space operation based on dimension and perception, so as to guide students to understand and feel the concept and connotation of modern building space at the beginning of design teaching.

Dimension: space with human as the dimension requires researching the physical relation between individual and space, object used, even the mental relation, with the purpose of improving situation of use according to the posture and move of human. Building and human body will actually meet on the level of ergonomics, "dimension of human" is a relatively vague but fully ideological word, relation of size, form and posture is its essence. Dimension first refers to the relative measurement relation between the building and human body, how to use dimension, can it be dominated and controlled physically (walk in, move around, stay inside) or mentally (accept as a symbol, an occasion or reproduction). In addition, for the reason of aesthetic or symbolization, human body structure is generally compared with building, the subjective reproduction of universe order uses human body instead of atom as the main reference point, uses the body as the base to compare size, geometry and amorphism, hard and soft, wide and narrow, strong and weak. The perception of beauty is always closely linked with the form of human body.

Perception: to an individual with five senses freely controlled, the experience in building is visual and dynamic first, then hearing, smell and touch, however, if building cannot meet one of these, the quality of this work will be greatly reduced. Five senses of human are not pure physiological functions, but also skills to learn, intelligence, learning and memorizing ability will link the operation of these senses and special experience, culture and time. Move of body can help human measure the size of object and space. Walking, visiting, dancing, arriving…these moves can make human better comprehend dimension, and explore some hidden part. Approaching, leaving, winding, climbing, accessing…these moves invite people into a set environment to independently organize things to see, hear, smell, taste and touch. In portrait and picture, building is only an image, once it is completed, it becomes various scenes to experience and move, even a series of scenario composed by feeling.

Space operation: philosopher and mathematician define space as constellation composed by object, such space is often located out of the ground where human is, located in the center of consciousness experience and abstract. Space must be hollow, limited from outside, filled inside, while building space is immaterial part defined by material, some part of the world is divided to adapt to living and use by human, and this is the essence of architectural design. A great breakthrough of modern architectural is that the building space is manifested to understand the space configuration. To architect, gap between ground, wall and ceiling can be anything but never nothingness; in fact, it is just the foundation for architect to take action, this hollowness is made into a container, it is provided with a clear form, shape and appearance of site are provided, as well as freedom to move. After modernism appears, building space has new meaning in operation. Organic character: connotation that local is affiliated to whole; mobility: different from close or open-flowing, through, partitioned but not separated space; implication: build the building space to realize symbolic effect.

3. Typical case assistance

Two typical small detached residences are selected in this course as cases to guide students to cognize and learn space, structure, flow line, dimension, etc. through making large-scale material model (1:50), drawing analytic drawing. Two cases are located in the downtown of Tokyo, and well tally with the setting of this course on site and function with good reference value.

(1) Dango Saka residence: designed by Kazuyo Sejima, with the land area of 24.3m^2, floor area of 50.4m^2. Wood frame structure system makes family life draw the outline

room, space is distinguished by fine difference of elevation, traditional partition wall is canceled, and internal space is richer and funnier.

2) Tsuchihashi residence: designed by Kazuyo Sejima, with the land area of $72m^2$, floor area of $10m^2$. Steel frame structure system, staggered flab opening creates a full-height atrium space, introduces sufficient light indoor, open concept contacts each floor visually, entering from mutually paralleled stair, people will feel the change of structure when moving in the space.

foresaid two typical cases imply three policies of modern space organization: structural space, raumplan and free plan. Structural space refers to the rigorous and consistent relation between the order of bearing structure and space graphic it generates, and the consistence of this research creates non-ambiguous clear organization structure. Raumplan is the word created by Adolf Loos and Josef Frank at the beginning of the 20th century, as a form of conceiving internal space, space height is decided according to the nature and purpose of place, structure is the external support, serving the internal atmosphere. Free plan is not anarchy of production nor denial of order, such space organization technology is developed in the first half of the 20th century, specially emphasizes on mutual penetration of space instead of concatenation, alignment or section stacking of room.

. Design content

The place of course design is selected in the urban historic block where students have conducted environment cognition survey previously, three representative plots are selected, relevant land and line scope is set, students can choose one to complete the design. The area of single plot is $0~70m^2$, single side or adjacent two sides facing the street, with 1~2 floor traditional dwelling around. The building function is small household detached residence, family members (a young couple) have occupational background of construction industry, with two underage children. The new floor area is not more than $100m^2$, the building height $\leq 8m$ (total height of internal usable space, excluding parapet wall, no underground space is set), which shall include living room, dining room, master bedroom, children room, multifunctional room (including working space and guest bed), kitchen, dining hall, toilet (1/2) and necessary storage space. In addition, parking and charging space for at least two electric bicycles shall be considered nearby the access of new building.

In teaching schedule, the course is divided into two stages. The first stage is typical case analysis (3 weeks), students experience and cognize two typical residence cases through large-scale material model on space, structure, flow line, dimension, etc. The second stage is design operation (5 weeks), students complete the whole course design task through professional plan, facade and section drawing, cooperating with material model of different scales and section perspective, analytic drawing, effect drawing, etc.

. Parallel thinking

Besides the teaching essentials set by the teaching plan, students are also required to have parallel thinking and research on the following four aspects, and finally reflect in an integrated design plan.

Place and interface: building form is the reflection of its internal space, while external space of building refers to the environment surrounding the building or environment between the buildings. Place externally limits the generation condition of building space, elements as building, road, greening, etc. in the place shall be comprehensively and rationally laid according to the current condition of base, building and other elements in the place are designed to form an organic whole to have function. External facade of building form, as the vertical interface to limit urban external space, directly reflects the building relation in the city, comfortable urban public open space also first depends on rational building form and layout.

Space and function: the most fundamental purpose of architectural design is to obtain usable space. The increasingly complex function requirement of modern building, breakthrough of construction technology and material provide architect with more possibilities to create building space, space meaning becomes the most important connotation of modern building, with meaning far more important than the usable room inside the building. Modern architectural design is a design of building space to a large extent, building form language and design method are themed in space. Building function means that space in and out of the building shall meet actual use requirement, which answers the question of basic purpose of building.

Flow line and access: flow line of building, known as dynamic line, refers to the moving route of pedestrian and vehicle in the building, various space is organized according to different behavior modes. On one hand, the internal functional space of building requires rational level, vertical traffic for mutual communication and contact; on the other hand, the internal space of building requires considering the rational connection with environmental condition surrounding the place, such as influence of continuity of street interface, selection of access position and setback treatment, exterior wall interface of surrounding building (including exterior window on) on new building, spacing between buildings and visual interference, rational use of daylighting, etc..

Dimension and perception: space inside the building is for use, therefore, the dimension of functional space in the building must use human body as the basic reference and consideration and combine with various behavior modes of human body to confirm rational building space dimension. Students shall pay attention to understand the relation between space element and space experience through graphic expression in space form treatment, mainly including sense of dimension and enclosure. Building dimension is also limited by construction condition, and has the reference relation with the environment.

6. Time node

First week: understand typical drawing data, make working model. Survey the place, make aplace model in group (base 60cm×60cm×5cm), use 1:50 material working mode to conceive the preliminary plan.

Second week: make a typical structure model, draw typical section perspective analytic drawing. Deepen field survey, deepen preliminary plan.

Third week: complete aformal model of typical case. Use 1:50 hand drawn plan, facade and section drawing, research function and space, flow line and dimension, through picture collage, perspective along the street to research relation and effect of place and interface.

Fourth week: confirm basic design plan, promote section, facade design, use working model as assistance (centralized drawing hanging for comment).

Fifth week: deepen design plan and specify design details, model and research internal space effect.

Sixth week: deepen 1:50 drawing, start making picture collage effect drawing (centralized drawing hanging for comment).

Seventh week: make 1:20 section perspective and analytic drawing, make 1:20 large-scale model (centralized drawing hanging for comment).

Eighth week: arrange drawing, typeset and complete course defense.

7. Result requirement

Two A1 grayscale drawings, one paper model (scale 1:20), several working models. Drawing content shall include:

(1) General plan (1:200), floor plans, horizontal and vertical section drawing and main facade drawing (1:50), one internal space organization section perspective drawing (1:20)

(2) Design specification and main technical and economic index (land area, floor area, plot ratio, building density).

(3) Express design intent and analytic drawing of design process (function analysis, flow line analysis, structure system, etc.).

(4) Paper model picture and computer effect drawing, picture collage, etc..

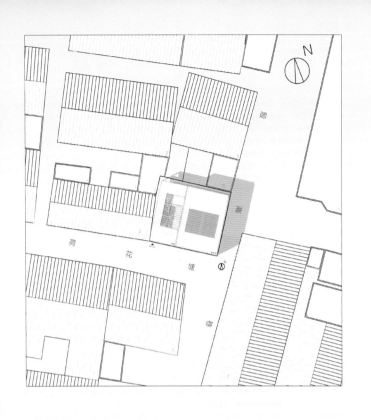
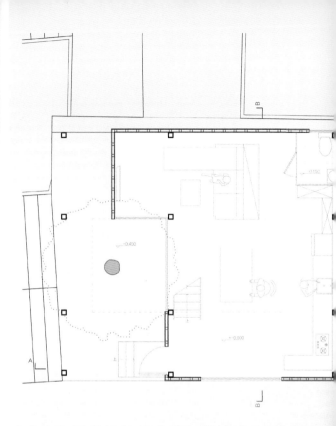
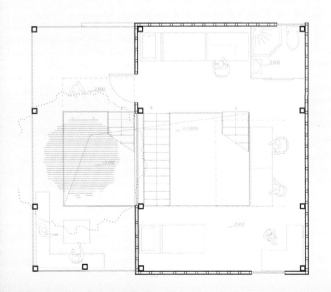
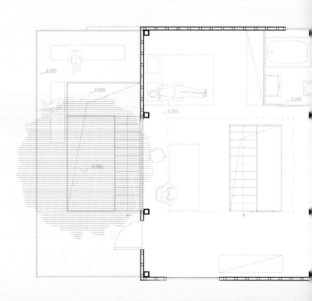

地从外部限定了建筑空间的生成条件,需要结合基地现状条件对场地内的建筑、道路、绿化等构成要素进行全面合理的布置,通过设计使场地中的建筑物与其他要素形成一个有机整体发挥效用。

ace externally limits the generation condition of building space, elements as building, road, greening, etc. in the place shall be comprehensively and rationally laid according to the current ondition of base, building and other elements in the place are designed to form an organic whole to have function.

以人为尺度的空间需要研究个人和他所使用的空间、物体之间的物理关系，甚至心理关系，目标是根据人的姿势和动作来改善使用的情况。
Space with human as the dimension requires researching the physical relation between individual and space, object used, even the mental relation, with the purpose of improving situation of use according to the posture and move of human.

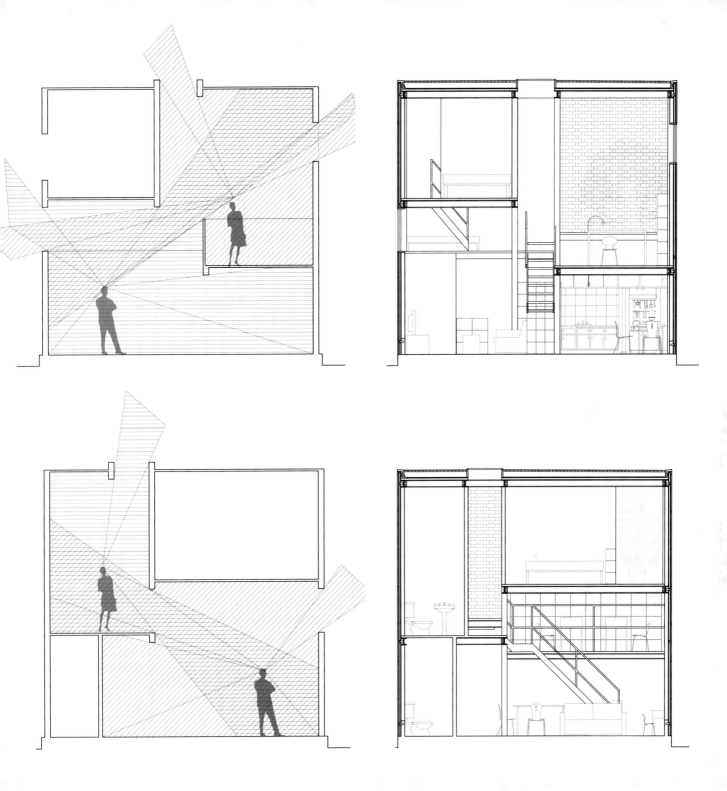

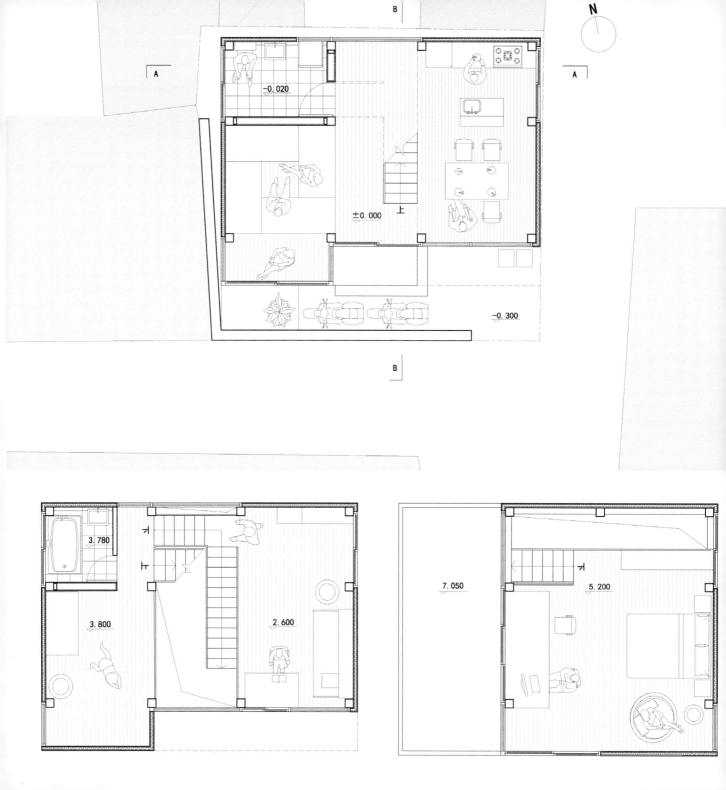

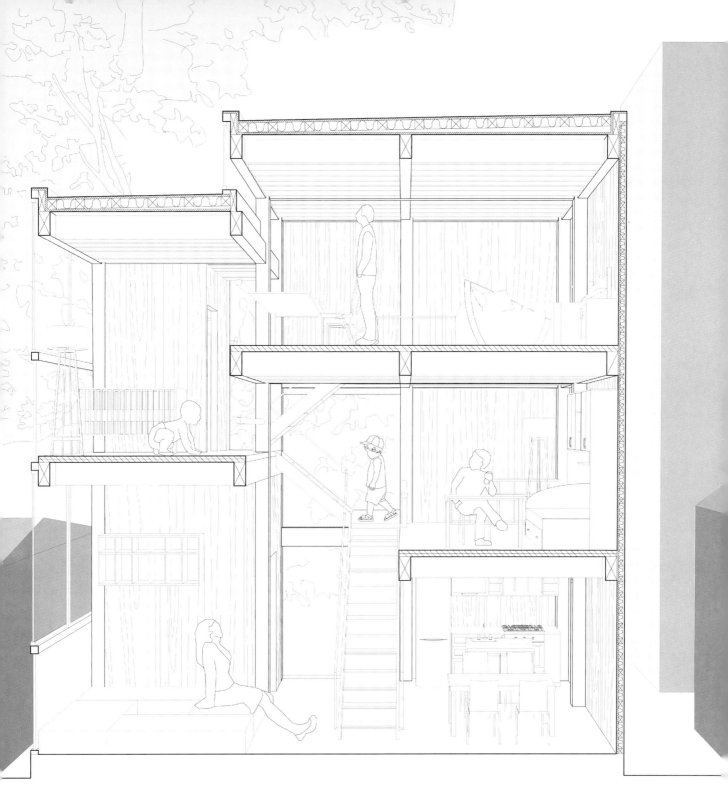

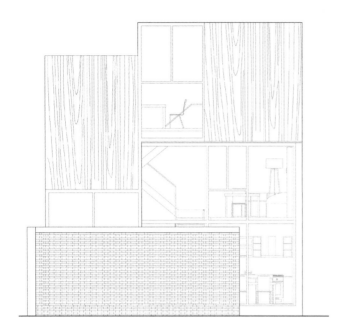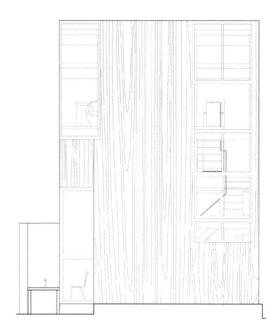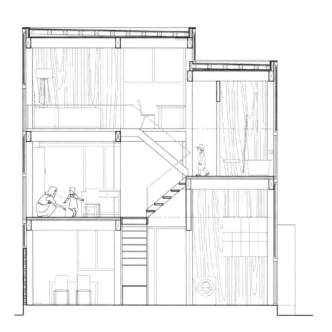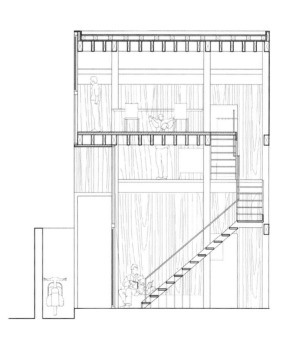

建筑设计（三） ARCHITECTURAL DESIGN 3
赛珍珠纪念馆扩建
EXPANSION OF PEARL BUCK MEMORIAL
周凌 童滋雨 窦平平

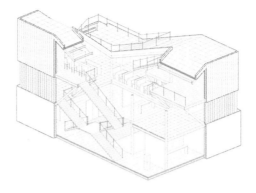

1. 目标

大三上学期的建筑设计课是本科阶段第三个综合设计。此课程训练目标设定为让学生掌握建筑设计中最基本的建造材料问题。通过这一建筑设计课程的训练，使学生在学习设计的初始阶段就知道房子如何造起来，深入认识形成建筑的基本条件：结构、材料、构造原理及其应用方法，同时课程也面对场地、环境和功能问题。训练核心是结构、材料、场地，在学习组织功能与场地的同时，强化认识建筑结构、建筑构件、建筑围护等实体要素。

2. 任务

课程设计题目为：赛珍珠纪念馆扩建。赛珍珠故居位于南京市鼓楼区的南京大学北园内，是美国作家赛珍珠在南京的故居。1919年起，赛珍珠和她的丈夫、农业经济学家约翰·洛辛·布克及家人居住于此，直至1934年。在这里度过的十多年间，她完成了处女作《放逐》和后来获得诺贝尔文学奖的小说《大地》等许多作品。2006年6月5日，南京赛珍珠故居被列为江苏省文物保护单位，2011年旧居被改造为赛珍珠纪念馆。为更好地展示历史，拟将旧居部分作为复原展示，完全还原当初的室内陈设，而纪念馆所需展示、导游与配套服务设施则安排在室外（包括地下）另行加建。计划室外扩建部分功能主要包含平面展示、多媒体展示、导游处、纪念品以及茶餐厅等服务休息设施。目前按规划要求，基地内地面最大可建面积约450m²。

设计总建筑面积约500~600m²，建筑层数地上主体结构1~2层。建筑限高：檐口高度不超过7m，总高不超过10m。地下层高不限。考虑与历史建筑和校园环境协调，尤其是与赛珍珠故居以及北侧二层历史建筑的总体关系。要充分考虑材料建造与实施的可能性。

3. 要求

设计初期，训练学生建立环境意识，要求关注基地周边的场地条件和文化特征。

关注一些要点，比如：（1）文脉：充分考虑校园环境、南北侧历史建筑、校园围墙以及现有绿化，需与环境取得良好关系。（2）退让：建筑基底与投影不可超过红线范围，如若与主体或相邻建筑连接，需满足防火规范；（3）边界：建筑与环境之间的界面协调，各户之间界面协调，基地分隔物（围墙或绿化等）不超出用地红线；（4）户外空间：扩建部分保持一定的户外空间，可以是地下空间；（5）地下空间：尽可能利用地下空间。

4. 功能

功能安排考虑都是开放空间，只有区域的划分，不要求一个一个固定的房间，功能可敞开布置，也可独立布置。具体分区为：展示区域（300m²）、导游处（25m²）、纪念品部（25m²）、茶餐厅（50m²）、厨房区域（≥10m²）、公共卫生间（1间）、门厅与交通等。

功能区的模糊设定，是鼓励同学灵活布置空间，让同学们可以更多样化地运用丰富的建筑语言，同时强调展览馆单层结构的完整性。

5. 材料建造

材料结构有预先准备的材料清单和结构选型。地下室部分直接要用混凝土墙或框架结构；地面部分可以选择钢、木、混凝土；围合与覆盖材料可以选择砖、瓦、木、石、土、金属、玻璃、塑料等。主要结构材料必须在指定材料中选择，其他料和辅材自定。结构类型以及相应的结构断面尺寸，会给出参考，比如钢筋混凝土架结构的跨度和梁高的关系，采用经验数据，即梁高=1/（8~12）跨度；钢结构跨度梁高的关系梁高=1/20跨度；砖混结构构造柱和圈梁给定尺寸240mm×240mm。这是设计中的关键节点，要求学生建立清晰的结构类型的概念，梁柱关系的逻辑要准而清楚。实现跨度是建筑设计的基本任务。结构训练阶段的表达方式是模型制作。

建立结构概念之后，墙体和覆盖的维护体系要进行选择和设计，要满足室内舒适度和热工性能要求，也要满足通风采光和视觉心理的需求。构造设计要配合1:20的墙身大样图。最后的和设计相关的特殊节点，学生可以自行选择表达，大样图比例自定。

6. 制图要求

作业最后规定4张A1图纸。必需的内容为：（1）总图与场地分析。特别要求必有一张大范围的基地位置图，表达周围城市肌理；总图除了表达楼层数、出入口、房间功能等基本信息以外，还要求表达设计建筑与周边相邻房屋的关系、场地高差、质感、树木等信息。（2）1:100的建筑方案平立剖面图，要求充分反映场地周边环境，周建筑、树木要求仔细刻画,写实表达。（3）1:50的平立剖各一张，放大表达材料厚度。（4）1:20墙身大样。（5）结构模型轴测图。（6）大比例建造分解轴侧图。其他图纸自由表达。

7. 时间进度

课程一共8周时间。由于课程设定的重点是材料建造，前期方案阶段时间不长，必须快速推进，鼓励学生用最简单的空间形态，使用单一的结构形式来生成覆盖，避免复杂的形体和多种结构混合，也避免前期过多的形式，把时间节约出来给后期的建构设计。第1~2周，训练主题为"型"，关注重点是场地、空间、形式。第1周完成基地调研、认知、分析，最快时间做出场地工作模型。第2周完成概念方案。第3~6周主题是"材"，关注围合空间的实体部分，分别进行结构、维护、覆盖构造的设计。具体来说，第3周进行结构设计（空间—结构）；第4周完成围护设计

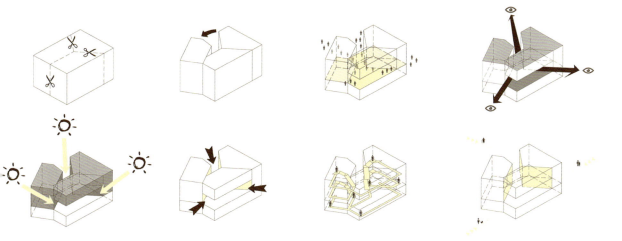

、开口、质感、保温）；第5周进行覆盖设计（屋顶、地面、基础、保温排水）；第6
进行细部设计（材料、细部、大样）。第7~8周制图、排版。

8.完成情况
多数同学能够按照课程设定的时间完成分阶段的训练。教师在前期方案阶段，帮
学生尽快做出方向选择，中期在结构构造等技术方面指导学生，后期在细部设计和
图方面提供部分参考图。在结构选型中当学生造型遇到结构矛盾时，需要回头修改
型，最终要使用最合理的结构来实现展览馆的空间覆盖，并且做出创新的形式。
整个题目设定，对于学生建立建筑设计中结构和空间的正确概念具有很大作用。
过设计训练，学生知道一个房子不只是排平面、做外形，而是一套空间、结构、构
的系统。这套系统不一定是由功能开始，利用结构、构造技术来实现功能平面并不
独立的阶段，学生应该一开始就有技术思考，从技术限制开始，从结构材料开始，
思考建筑的空间和功能。最终的设计应该是一套自我完善的技术体系。

Objective
chitectural Design in the first semester of year 3 is the third comprehensive design
the undergraduate stage. The training objective of this course is to make students
ster the fundamental construction material in the architectural design. In this
urse, students are trained to know how a house is built at the beginning of learning
sign, and deeply understand the basic conditions for forming architecture: structure,
terial, building principle and application method, and the course also aims at place,
vironment and function. The training core includes structure, material, place, while
rning organization function and place, students can strengthen the understanding
substantial elements as architectural structure, architectural component,
hitectural enclosure, etc..

ask
e course design subject is: Expansion of Pearl Buck Memorial. Former Residence
Pearl Buck is in the northern campus of Nanjing University in Gulou District,
njing, as the former residence of American writer Pearl S. Buck in Nanjing. Since
19, Pearl Buck and her husband, agricultural economist John Lossing Buck and
ir family had lived here, till 1934. During that period, she completed her maiden
rk The Exile and novel The Good Earth that won Nobel Prize in Literature later. On

June 5, 2006, Former Residence of Pearl Buck was listed as cultural relics protection unit in Jiangsu Province, in 2011, Former Residence of Pearl Buck was reformed into Pearl Buck Memorial. To better display the history, it is planned to use the former residence part as recovery display, completely recover original interior furnishing, and further expand the display, guide and supporting service facility of the memorial outdoor (including underground). According to the plan, the function of outdoor expansion mainly includes graphic expression, multimedia display, information area, shopping area, tea area and other service resting facility. At present, according to the planning requirement, the maximum buildable area on the ground in the base is about 450m^2.

The total design floor area is about 500~600m^2, aboveground main structure has 1~2 floors. Building height limit: the cornice height is not more than 7m, the total height is not more than 10m. Underground floor height is not limited. Considering the coordination between historical architecture and campus environment, especially the overall relationship between the Former Residence of Pearl Buck and two-floor historical architecture in the north, the possibility of material construction and implementation shall be fully considered.

3.Requirement
At the beginning of the design, students are trained to establish the awareness of environment, concern the place condition and cultural characteristics surrounding the base, and concern some essentials, such as: (1) context: fully consider the campus environment, historical architecture in the south and in the north, campus fencing wall and existing greening, obtain good relationship with the environment; (2) retreat Distance: building base and projection cannot exceed the red line. If connected with main body or adjacent building, shall meet fire protection code; (3) boundary: interface between building and environment is coordinated, interface between households is coordinated, base partition (fencing wall or greening, etc.) cannot exceed the land red line; (4) outdoor space: the expanded part maintains certain outdoor space, which can be underground. (5) underground space: properly utilize underground space.

4.Function
Separate fixed room is not required, the space is open with only area division, the

function can be laid open or separate. Specific division: exhibition area (300m^2), information area (25m^2), shopping area (25m^2), tea area (50m^2), cooking area ($\geqslant$10 m^2), washroom (1), lobby, corridor, etc..

The vague setting of functional area can encourage students to flexibly set space, and apply rich architectural language in a more diversified way. At the same time, it emphasizes the integrity of single layer structure of the exhibition hall.

5. Material building

Material structure has pre-prepared material list and structure type selection. It is directly required to adopt concrete shear wall or frame structure for the basement part; steel, wood, concrete can be selected for the ground part; brick, tile, wood, stone, earth, metal, glass, plastic, etc. can be selected for enclosing and covering material. Main structure material must be selected from the designated material, other material and auxiliary material can be chosen by students. There will be a reference for structure type and relevant structure cross section dimension, for example, span of reinforced concrete frame structure and relationship with bean height, empirical data is adopted, that is, beam height=1/(8~12) span; relationship between steel structure span and beam height, beam height=1/20 span; provided dimension of masonry-concrete structure structural column and ring beam 240mm×240mm. Structure is the key node in the design, students are required to establish clear concept of structure type, and the logic of relationship between beam and column shall be precise and clear. Realizing span is the basic task of architectural design. The expression form at the structure training stage is model making.

After the concept of structure is established, the maintenance system of wall and covering shall be selected and designed to meet interior comfort and thermal performance requirement, and meet the requirement on ventilation, daylighting and visual psychology. Structure design shall cooperate with 1:20 wall detailed drawing. The final and design related special node can be independently selected and expressed by students, and the scale of detailed drawing can be chosen by students.

6. Drawing requirement

The regulation on homework is four A1 drawings. The required content includes: (1) General drawing and place analysis. A large-scale base location map is required to express surrounding urban tissue; besides showing the basic information as floors, access house function, etc., the general drawing shall also show the relationship between the designed architecture and surrounding adjacent house, place height difference, texture, trees, etc. (2) 1:100 architectural plan, facade and section drawing, which shall fully reflect the surrounding environment, surrounding building, trees shall be carefully drawn, expressed in realistic style. (3) 1:50 plan, facade and section drawing, which shall amplify the expression of material thickness. (4) 1:20 wall detailed drawing. (5) Structure model axonometric drawing. (6) Large-scale building break-down axonometric drawing. Other drawing is freely expressed.

7. Schedule

The course lasts for 8 weeks. The focus of the course is material building, the ea plan stage should not be too long and must be rapidly promoted. Students a encouraged to use the simplest space form, use asingle structure form to genera space covering, avoid a complex form and the mixing of various structures, avoid t many forms in the early period, and save time for the later building structure desig In the 1st and 2nd weeks, the training theme is "form", the focus is place, spac form. In the first week, students shall complete base survey, cognition, analysis, a make place working model as soon as possible. In the second week, students sh complete conceptual design plan. The theme of the 3rd~6th weeks is "material", a the focus is substantial part of enclosing space. Students shall respectively desi structure, maintenance, covering, construction. To be specific, students shall ha structure design in the third week; complete enclosing design (enclosure, openin texture, insulation) in the 4th week; students shall have covering design (roof, groun foundation, insulation drainage) in the 5th week; students shall have detail desi (material, detail, detailed drawing) in the 6th week, and Drawing, layout in the 7th~8 weeks.

8. Completion situation

Most students can complete stage training according to the set schedule. Teache help students make direction choice at the early plan stage, guide students at m term structure building, etc., and provide part of reference drawing at the late sta of detail design and drawing. When students have structure contradiction in structu type selection, students shall change the modeling, and finally use the most ratior structure to realize the space covering of the exhibition hall, and make innovati form.

The whole subject setting has great function on helping students to establish corre concept of structure and space in architectural design. Through the design trainir students know that a house is not only plane layout and appearance making, bu set of system of space, structure and building. This set of system may not start w function. Using structure, construction technology to realize functional plane is n a separated stage, students should have technical thinking at the beginning, fro technical restriction. Structure material, students shall think about space and functi of architecture. The final design should be a set of self-improved technical system.

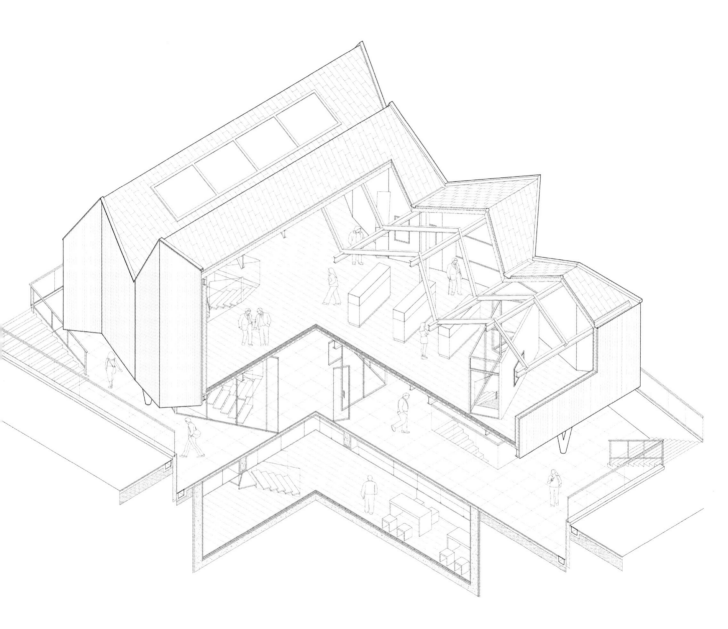

抓住场地要素，从流线、景观、地形等的限制中获取促成方案的元素，使建筑本身与赛珍珠纪念馆从属明确而又互相呼应。
Grasp the site element, obtain element from restriction of flow line, landscape, landform, etc. to promote the plan, make the building have clear affiliation and mutually echo with Pearl Buck Memorial Museum.

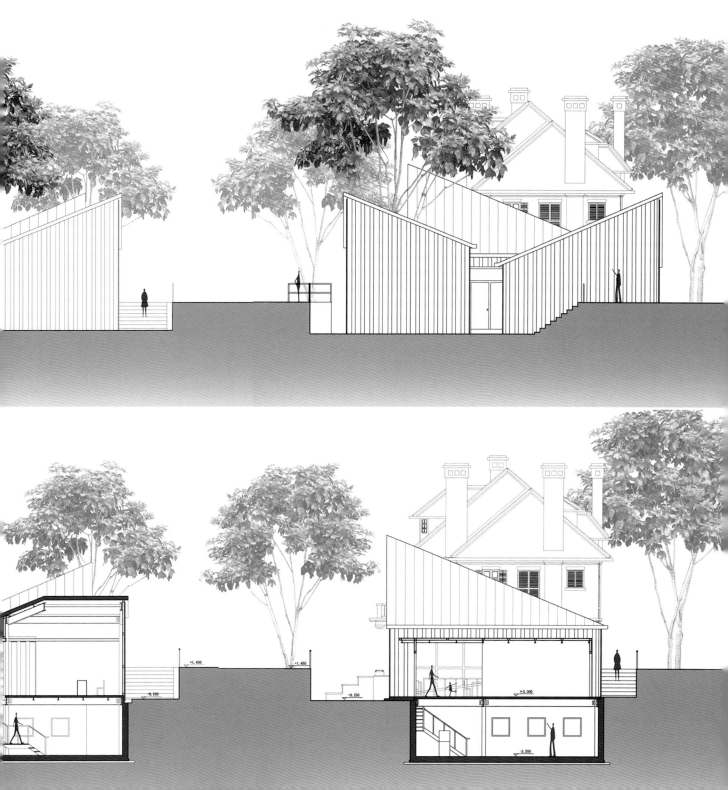

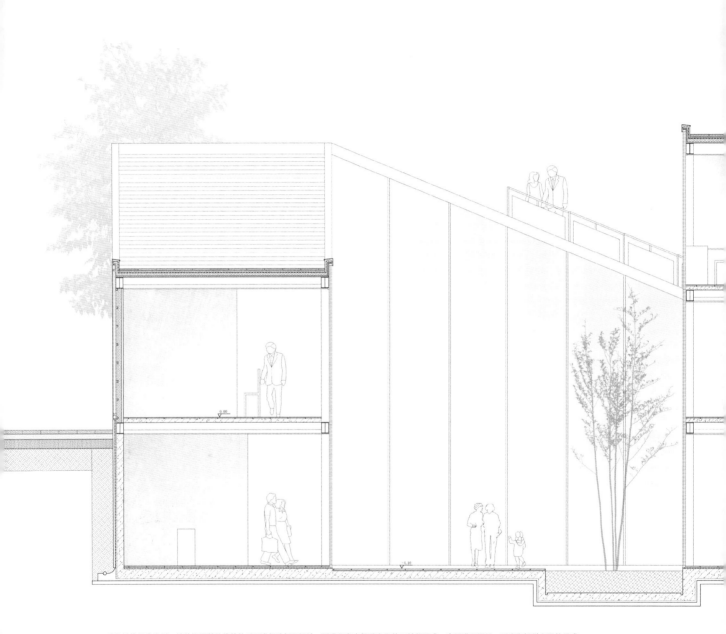

建立结构概念之后,墙体和覆盖的维护体系要进行选择和设计,要满足室内舒适度和热工性能要求,也要满足通风、采光和视觉心理的需求。
After the concept of structure is established, the maintenance system of wall and covering shall be selected and designed to meet interior comfort and thermal performance requirement, and meet the requirement on ventilation, daylighting and visual psychology.

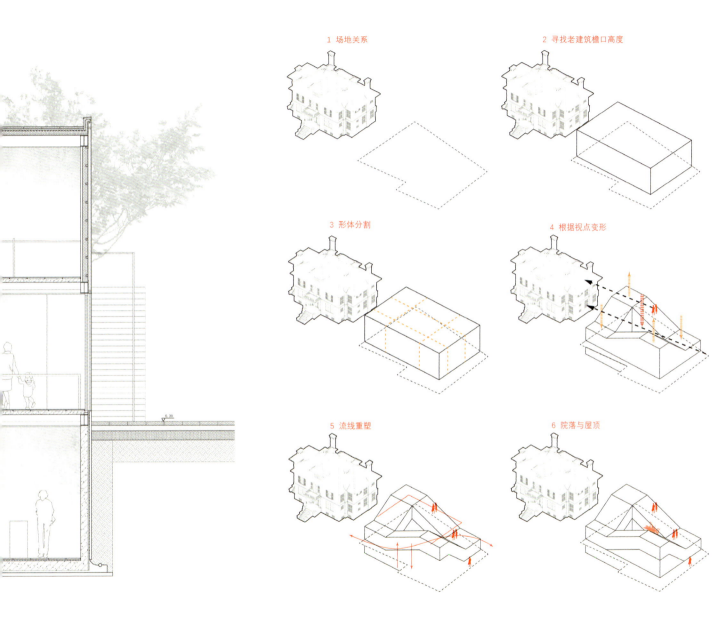

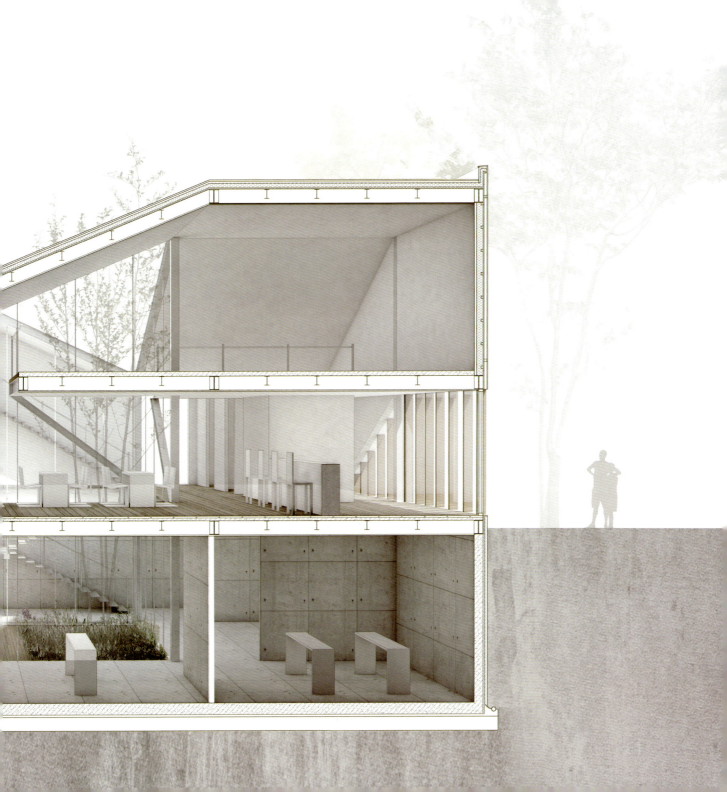

建筑设计（五+六）ARCHITECTURAL DESIGN 5 & 6
城市建筑：社区中心
URBAN ARCHITECTURE: COMMUNITY CENTER

华晓宁 钟华颖 王铠

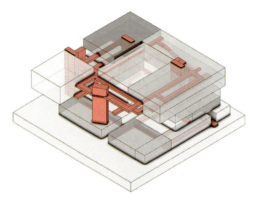

　　本学期三年级下学期建筑设计课程依然以"城市建筑"作为主题，通过中大型公共建筑的设计课题，培养学生在复杂环境中综合分析和解决建筑问题的能力。在课题设定和教学过程中，着重在两个方面进行了思考和探索。

　　建筑设计中城市性的彰显。何为建筑的城市性？这是一个当前受到广泛关注和讨论，但却又莫衷一是、众说纷纭的话题。我们认为，对于城市中的建筑而言，城市主要从物质性和非物质性两个方面对其产生制约和影响。城市物质空间系统是建筑存在的背景结构，也是城市生活的承载者。我们从城市物质空间系统的形态特性中提炼出一组关键词："实与空、层与流、内与外、轴与界"。

　　城市，是建筑与建筑师最重要的舞台。从物质环境上来看，它是由多样化的建筑实体及其之间的空间共同构成的，承载和容纳了城市居民纷繁复杂的生活。城市中各种物质、能量、信息乃至城市本身永远都处在不停息的流动和演化之中。

　　当代纷繁而多元的城市生活产生了错综复杂的城市物质空间系统，它给予城市中的建筑诸多限定。而城市中的建筑一旦形成，既同时介入和重构了城市物质空间环境，又改变和重新定义了城市生活本身。在某种程度上，"'事物之间'的形式比事物本身的形式更重要"。

In the second semester of the third grade, architectural design course is still theme in "Urban archite", of medium and large public building, the student's ability comprehensively analyze and solve architecture problem in complex environment cultivated. In the task setting and teaching, two aspects are emphasized for thinkir and exploring.

Manifest urbanism in the architectural design. Urbanism of architecture is a wide concerned and discussed topic, and there is not a decisive answer. We believe tha to the architecture in city, city restricts and affects architecture from materiality ar non-materiality. City material space is the background structure for the existence architecture, and also the carrier of city life. We distill a group of keywords from th special form of city material space system: "solid and empty, layer and flow, insic and outside, axis and boundary".

City is the most important stage for architecture and architect. From materi environment, it consists of diversified architectural body and space, carries ar contains numerous and complicated lives of urban residents. Various material energies, information in the city and even the city itself are always in permanent flo and evolution.

When the complicated and diversified urban life produces complex city material spac system, it limits architecture in the city on many aspects. Once the architecture in th city is formed, the city material space environment is intervened and reconstructe and city life is changed and redefined at the same time. To some extent, "for between things is more important than the form of things".

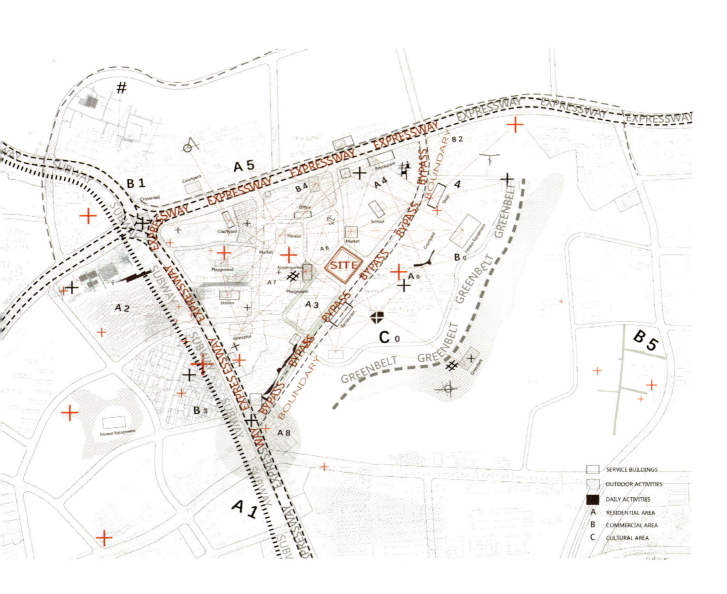

场地区位研究/Site location research

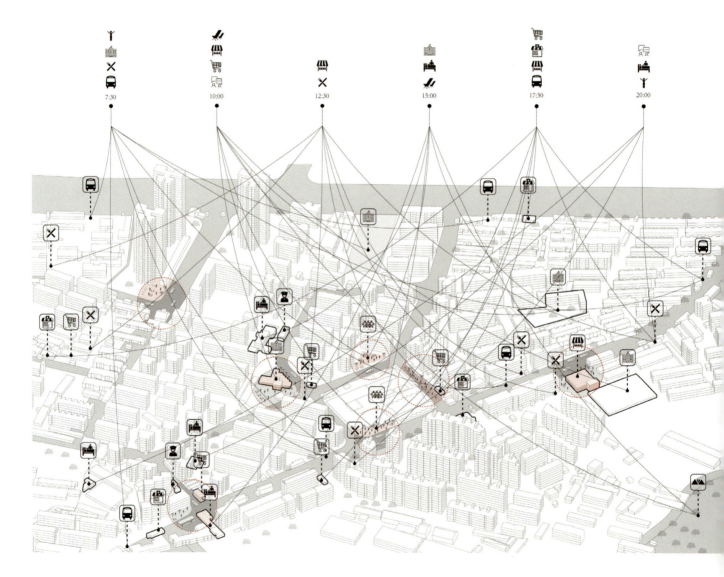

场地生活行为研究/Site living behavior research

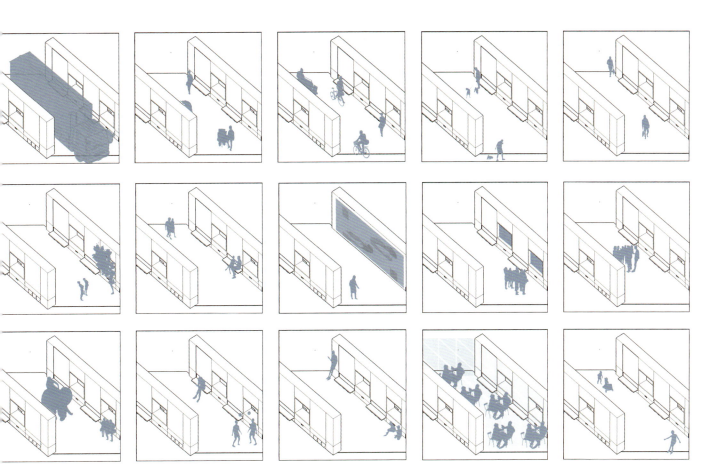

场地周边街道断面研究/Site surrounding street cross section research

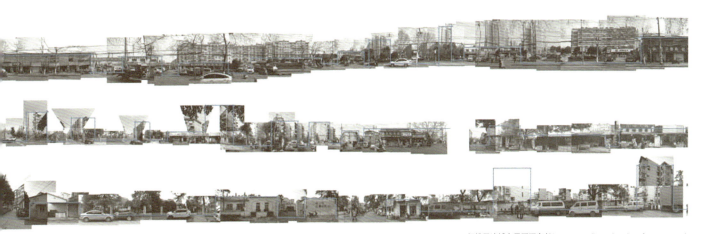

场地周边城市界面研究/Site surrounding urban interface research

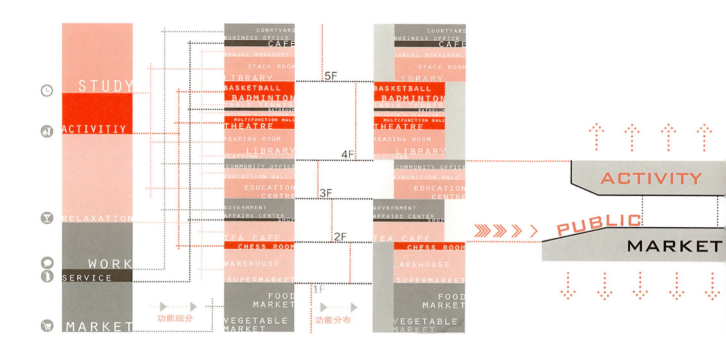

空间计划研究/Space plan research

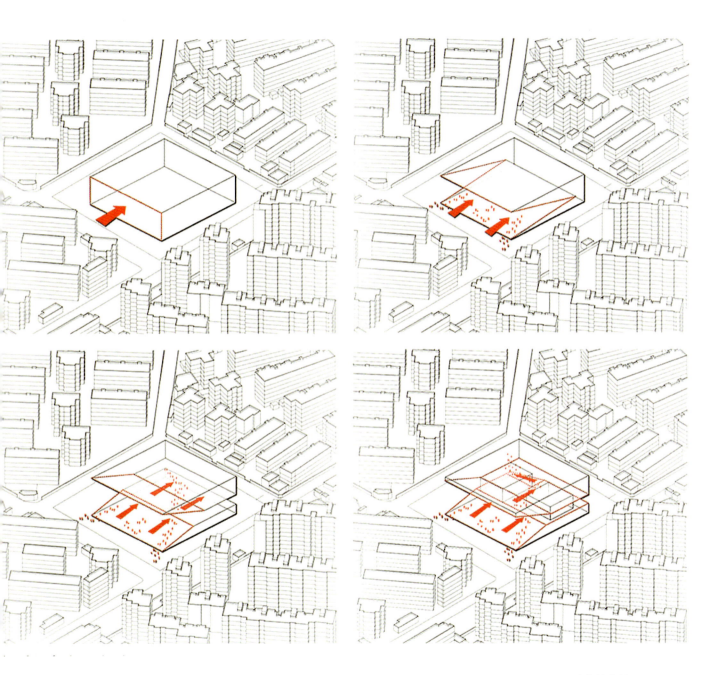

初始空间策略/Initial space policy

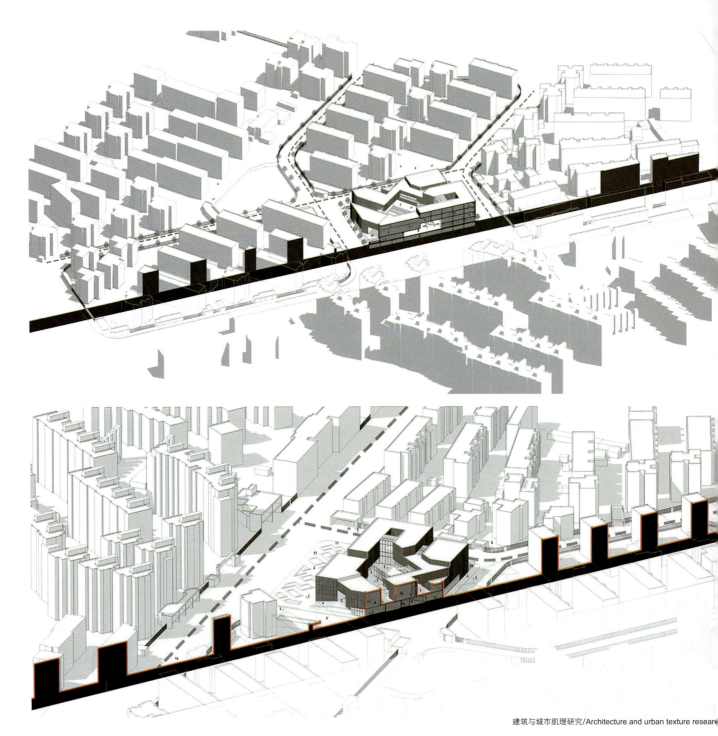

建筑与城市肌理研究/Architecture and urban texture research

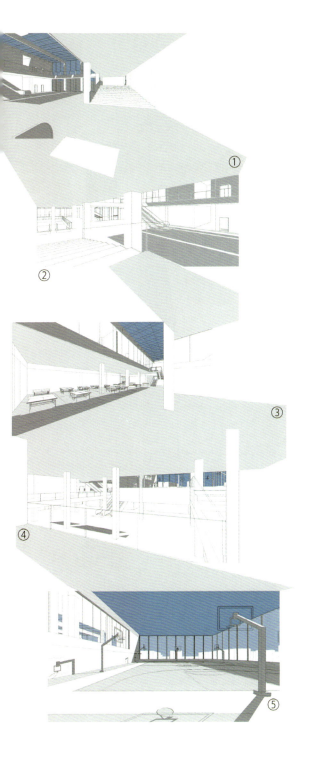
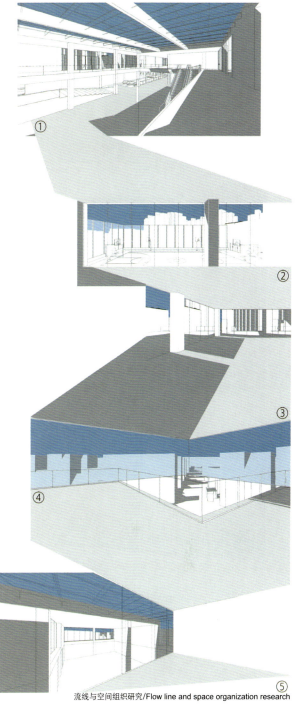

流线与空间组织研究/Flow line and space organization research

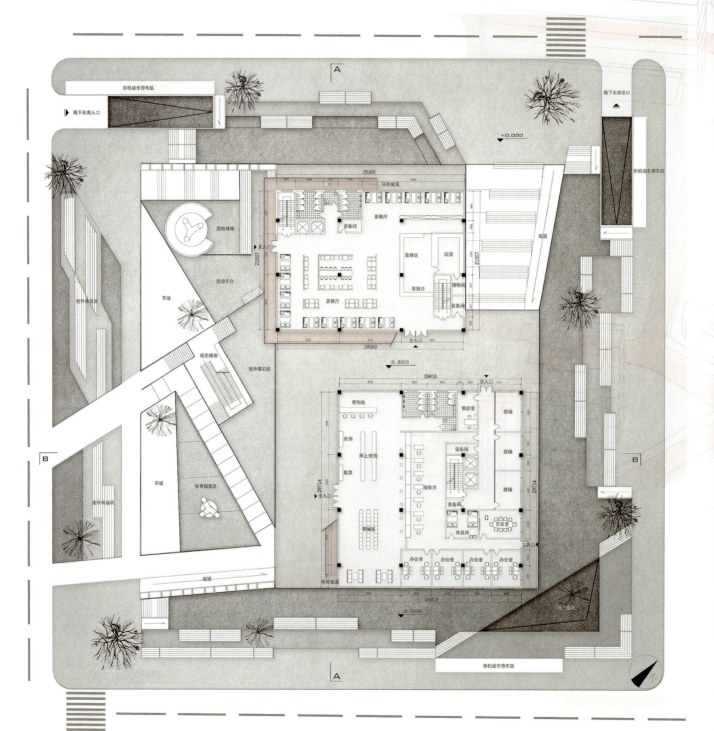

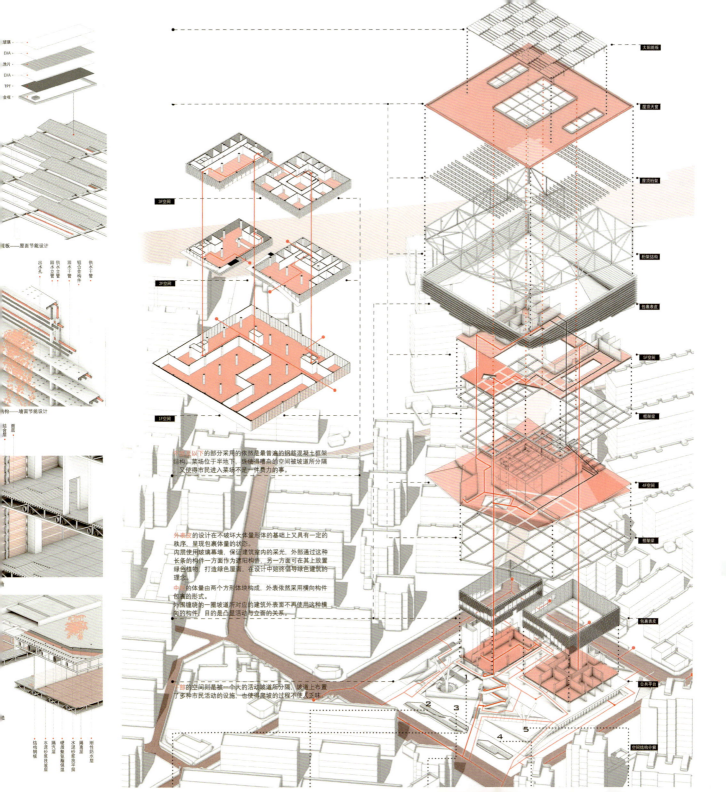

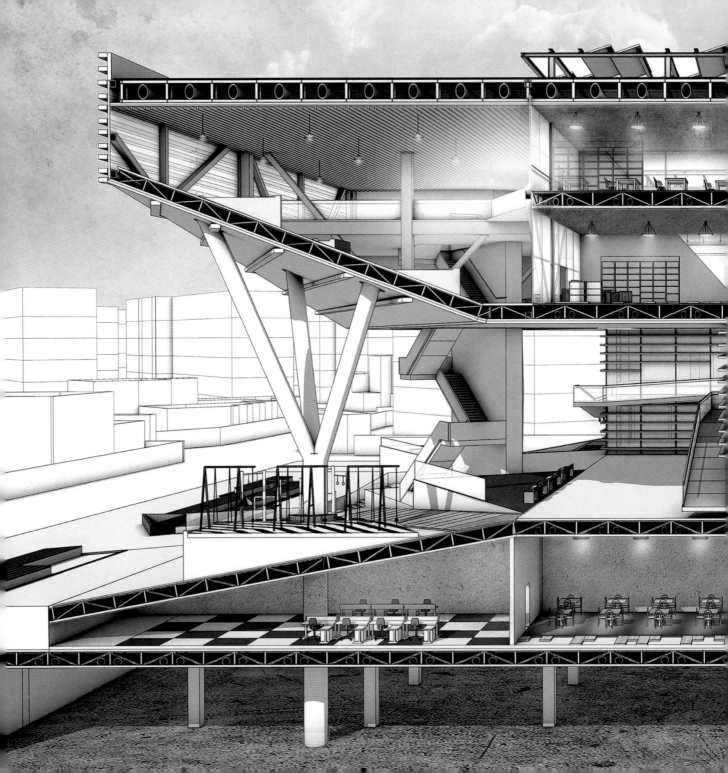

本科毕业设计 GRADUATION PROJECT
长汀历史名城更新与建筑设计
RENEWAL AND ARCHITECTURAL DESIGN OF CHANGTING HISTORICAL CITY

丁沃沃

城市化进程不仅改变了城市的面貌与市民的生活环境，也改变了建筑物的设计标准和设计方法。在城市中，建筑单体的造型不再仅仅关注自身的形象，更重要的是关注与其相关的外部空间。传统城市的环境已经是人们的主要生活环境，城市空间已经是人们主要活动的外部空间。为此，建筑设计教学应该在建筑设计中融入更多的城市空间要素，建立建筑与空间的关系，强化空间训练，强调建筑内部空间与外部空间的融合。为更好地体现设计训练的目标，本课题依然延续了去年的设计场地，结合老城更新的实际项目，将研究与设计相结合，探索研究性设计的方法与路径。

在去年研究的基础上，本次设计训练选择两块急需更新的城市地块，通过对历史资料的研习和对老城空间的调研，分析传统城市空间的基本特征和建筑语言。更为重要的是，学生们通过对传统建筑的切身体验理解建筑文化的魅力，探索在城市更新的过程中延续传统意向的方法。通过课程训练增强学生的城市意识，开阔建筑学的视野，掌握设计研究的方法。

福建长汀是国家历史文化名城，在历史文化名城过度开发的今天，长汀老城却并没有得到应有的开发和更新。老城设施陈旧，公共活动空间稀少，旅游设施完全不能支撑老城旅游业的发展。长期以来，地方政府一直想开发旅游业。为打造长汀的老城形象，政府拆除了部分形象或质量不佳的老街区，同时动用大量资金从外地购置传统民居或仿古建筑整体安置在这些地块之中。然而此举不但没有增加长汀的城市特色，却招来学界的种种非议。所以，结合老城发展的方向，从提升城市公共空间的角度出发，探索城市设计的新方法，获得城市空间和建筑设计两者间的平衡是本轮城市设计的重点。

长汀老城北依卧龙山，东伴汀江水。城中，街网密布，院落相连。就长汀老城的特征而言，有趣的是街巷和院落，缺乏的是公共活动空间，因此，利用街巷和院落打造现代长汀人和旅游者都期盼的城市公共空间成为本轮毕业设计主要研讨的课题。根据城市设计提供的资料，我们在老城里选择了两个地块：

A地块北临长汀老城贯通东西的主要街道新民街，新民街是目前旅游者的必经之地，也是当地政府主要装扮的街道。B地块内现在的街道卫生院准备搬迁，为城市的更新腾出空间。由于地理空间的位置很好，在与当地政府的协商研究之后，拟建一组以酒吧或水吧、表演等空间为主的特色商业建筑群。

B地块邻近长汀老城的中轴线——南大街，周边人流密集。A地块所在的街区中间被一街道工厂占用，街区内部的道路也就隆之湮灭。工厂的搬迁给城市更新带来了契机，内部遗存的传统建筑也给城市更新留下了记忆的痕迹。通过调研，同学们与工厂搬迁之前的街区居委会接触，了解了这一历史街区大量的信息。通过和城市规划部门的沟通，结合城市旅游设施的需要，一个具有地方特色的餐饮建筑群的项目构思被落实。在任务明确以后，同学们主要研究了长汀城市物质空间的几何特征、景观特征、光影特征、人流行为特征、休闲场所特征等，尤其对城市的街巷和院落特征进行研究，奠定设计的基础。

设计概念是设计操作的目标，设计概念反映了设计者对设计任务的理解和解题视角。六位同学分别从城市空间和人流行为的角度出发，给出了自己的构思：

暮色晨光
方案以光为概念。在长汀老城调研中发现了多种不同的光进入方式，希望通过光的空间的归纳变形，提供给人以光的韵律的酒吧体验。

折巷
方案以折巷为概念，从长汀老城狭窄而曲折的传统街道空间出发，希望将这种趣的街道空间延伸到街廓内部，同时充分利用院落，创造有传统韵味的现代酒吧街。

共享
方案以共享为概念，对传统空间类型进行分类，通过对传统空间的分析、转译现空间的共享，同时在打通的方向上创造空间的流动性，设计出丰富有趣的空间体验

景廊
方案以视线为概念，充分利用长汀老城之中可以看到的所有景色，希望人在其行走、吃饭的时候可以感受到长汀的特色。

沿街美食
方案以沿街吃为概念，以街道加餐桌为基本尺寸在场地中进行缠绕，希望在这可以为旅游提供团餐的场地中创造不同的餐饮体验。

光满厅堂
方案选取城市中居民对光的应用作为体现城市特点的核心概念，将光与饮食结合起来，让人体会在不同的光环境中吃饭的乐趣。

尽管以上构思丰富多彩，但都有共同的特点：在场地设计上融入城市空间，疏被打断的街巷，增加城市街区的通达性，融通室内外空间，增加城市公共活动空间数量和面积。在建筑设计上，探索空间的传承与创新，探索传统建筑文化的延续与变异探索以建筑界面作为主体的材质设计手法。最终每一个方案都给出了既有传统韵味有现代空间气息的丰富的建筑空间和城市空间。

通过设计工作坊，同学们理解了城市建筑的内涵，懂得了城市对建筑的重要意义他们每个人都制作了一个小影片，以动态影像的方式表达了学习的成果。

banization has not only changed the appearance of cities and the living environment of citizens, but also brought new requirements on the design standards and design methods of architectures in the city. The shape of an single architecture in the city shall take into consideration not only its own image but also its external space. The urban environment has become the main living environment of citizens, to consider the urban space as the main outdoor space of their activities. Therefore, the teaching of architectural design should be innovated to contain more elements of urban space, including the relationship between architecture and space, intensified training on sense of space, the integration of the interior and exterior space of architectures. Continuing to combine with the real project of renewing the old city on the design site of last year, this project has made some exploration on the methods of design research based on the combination of research and design, so as to achieve the goal of design training.

Based on the research of last year, two pieces of plots of the old city under renewal be chosen for the design training. By carefully studying the historical data and surveying the space of the old city, the students have found out the characteristics of the space of the old city and the trait of the architectures in the old city. What's more, the students can experience the architectural culture through surveying the old city, and therefore to explore how to continue the spirit of the architectural culture in the renewal of the old city. Through training of design, students can enhance their urban consciousness, broaden their vision of architecture, and master the methods of researching in design.

While lots of historical and cultural cities of China are over-developed at present, Changting of Fujian, another historical and cultural city of China has not been properly developed and updated yet. Having obsolete travelling facilities and rare urban space for public activities, the old city of Changting can not catch up with the development in tourism. For a long time, the local government has been trying to build a new image of the old city of Changting, and develop its tourism. The government has removed parts of the old neighborhoods which had poor image and inferior quality, and meanwhile has spent a lot of money in purchasing lots of traditional mansions or antique architectures to fill the removed neighborhoods. However, this move has no effect on improving its urban characteristics, and in turn, incurs a lot of academic criticism. As a result, it's of great importance in this round of urban design to find a new method in urban design by increasing the urban space based on the old city's development which can achieve a balance between urban space and architectural design.

The old city of Changting is located to the south of Wolong Mountain and the west of Tingjiang River, being dotted with dense lane nets and continuous courtyards. The old city is rich in fantastic lanes and dedicate courtyards, but short of public urban space. Therefore, it has been the main project of this round graduation design to build new public urban space combining the trait of lanes and courtyards which can be benefited by both the modern citizens of Changting and tourists. According to the information provided by the urban plan, we have selected two plots in the old city.

Plot A is to the south of Xinmin Street, which is a main street stretching from the west to the east in the old city of Changting. As the most popular destination of tourists, Xinmin Street is one of the streets which are delicately renewed by the local government. The current neighborhood health center within Plot B is being moved out, releasing more space for the renewal of the old city. Embracing a perfect geographical location, Plot B will be built into an commercial complex characterized by space for bars and performance after coordinating with the local government.

Plot B is adjacent to the central axis of the old city of Changting, South Main Street, where is crowded with people. Formerly the center of Plot A was occupied by a neighborhood factory, which destroyed the inner streets inside the neighborhood. The removal of the factory has brought opportunities for the urban renewal, while the remaining traditional architectures reflect the historical memory of the old city. By investigation, the students got in touch with the former neighborhood committee before the removal of the factory, and discovered numerous information about the history of that street. Through communication with the urban planning authority, Plot A will be built into an architectural complex characterized by local foods based on the demand for urban travelling facilities. Under the explicit tasks, the students have separately done careful research on the traits of public space in the old city of Changting on the aspects of geometry profiles, landscapes, lights, crowds and their activities, leisure facilities, etc.. Especially, their study on the feature of lanes and courtyards of the old city has laid the foundation of design.

As the goal of design operation, design concept reflects the understanding about the design task of the designer and his/her design perspective. Six students respectively wrote out their design schemes from different views of urban space and crowds activities:

Day and night in fragmented light

This scheme is based on the concept of lights. A variety of light entering methods were found in the survey about the old city of Changting. This scheme is intended to reflect the summarizing and transformation of lights in space, providing a light rhythm bar experience.

Meandering lanes

The concept is based on meandering street, starting from the traditional street space of Changting which is narrow and winding, hope to extends the interesting street space into the district, and at the same time make full use of the courtyard, creating a modern bar street with traditional flavor.

Sharing

This scheme is based on the concept of sharing. By means of classifying traditional space categories, this scheme is intended to achieve space sharing on transferring the traditional space based on analyzing, and supplies a bunch of interesting space experiences by increasing the liquidity in the open direction.

View corridor

This scheme is based on the concept of sight. By making full use of all the scenes of the old city in Changting, this scheme will allow the people who are walking or eating in the view corridor to feel the features of Changting.

Foods alongside the street

This scheme is based on the concept of foods alongside the street. By placing streets full of dining tables and chairs surrounding the landscape sites, this scheme is intended to provide different dining experiences for tourists groups.

Lighting maximum

Taking the local citizens' application of lights as the core concepts to reflect the characteristics of the city, this scheme is intended to provide joyful dining experiences in different lights by combining lights and foods.

Through these above colourful schemes, we can draw a conclusion of a common feature of these schemes: they are all trying to design sites by taking urban space into consideration, to increase the accessibility of the urban streets by dredging the blocked lanes, and to increase the quantity and area of public urban space by integrating indoor and outdoor space. They have explored the inheritance and innovation of space and the continuity and variation of traditional architectural culture in the design of architecture, and found a new way to design the architectural facade as the main body of architectural material. In the end, each scheme has offered abundant architectural space and urban space which is combined with traditional charm and contemporary breath of space.

By doing research in the design studio, the students have understood the connotation of urban architecture and the importance of the city to the architecture. Each of them has made a short film that records their results of the study in a dynamic image way.

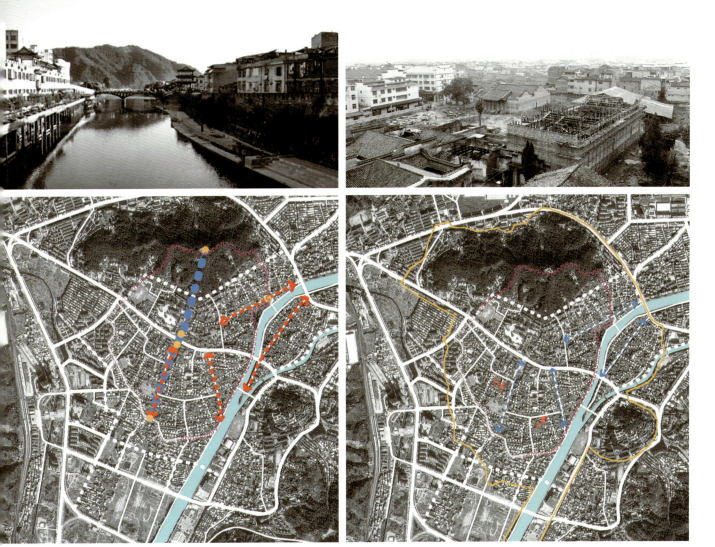

学生首先调研长汀老城的总体空间构架、交通体系和风貌特征，获得对整个区域的宏观认知。
Students first survey the general space structure, traffic system and style feature of Changting old city to obtain the macro-cognition of the whole region.

对长汀城中两个具有代表性地块的区位关系、交通可达性和产权分析等因素进行分析，初步提出各区域存在的主要问题。

Analyze location relation, traffic accessibility and property right analysis, etc. of two representative plots in Changting city, preliminarily propose main problems in each region.

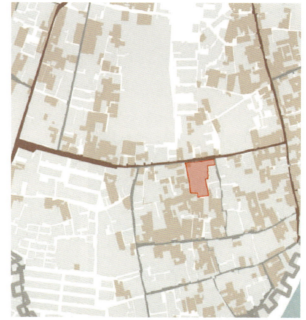

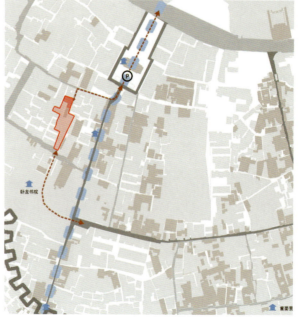

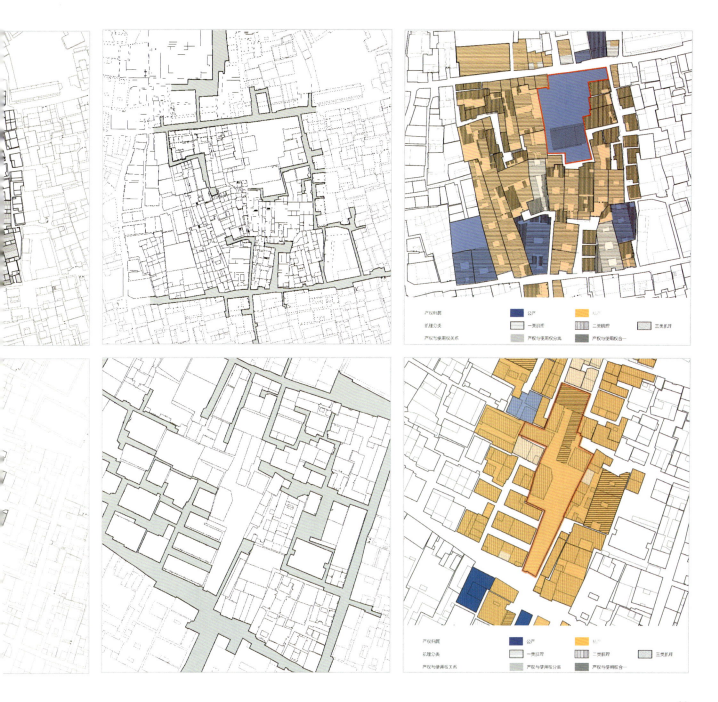

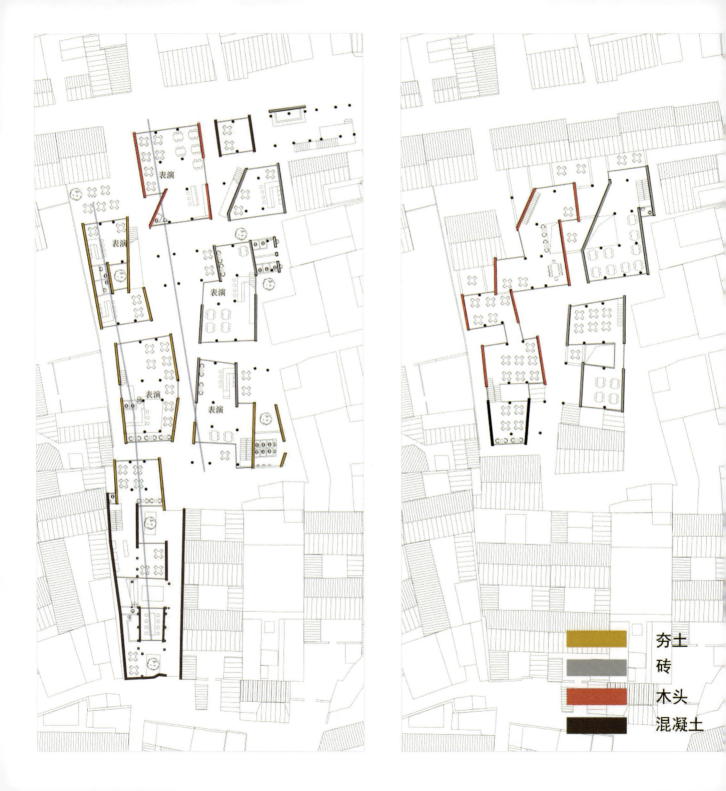

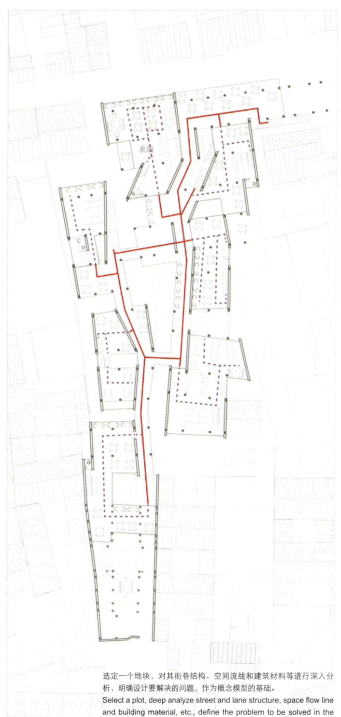

选定一个地块，对其街巷结构、空间流线和建筑材料等进行深入分析，明确设计要解决的问题，作为概念模型的基础。
Select a plot, deep analyze street and lane structure, space flow line and building material, etc., define the problem to be solved in the design, as the foundation of conceptual model.

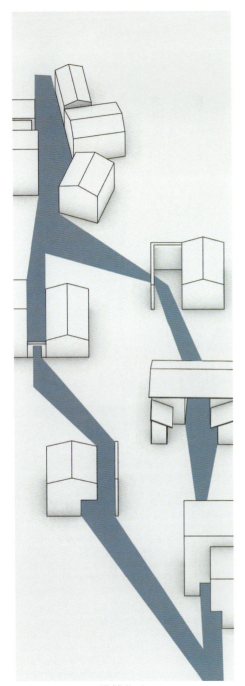

"共享" Sharing

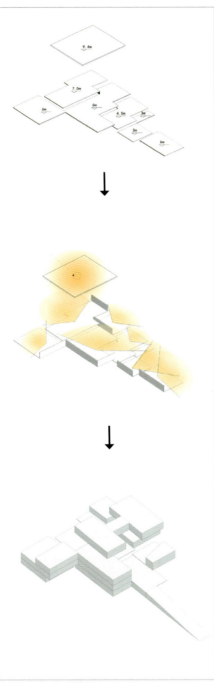

"景廊" View Corridor

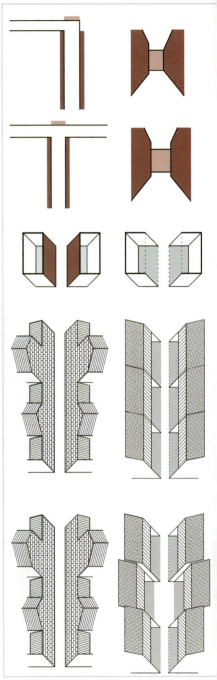

"拆巷" Meandering Street

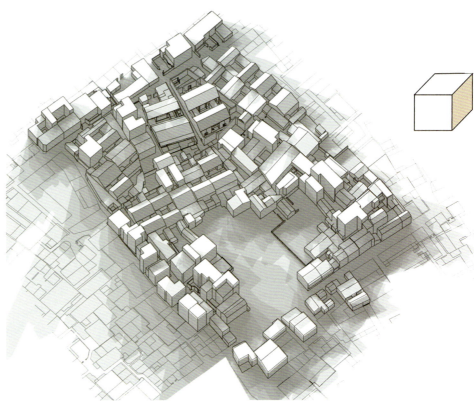

"暮色晨光" Day and Night in Fragmented Light

"沿街美食" Eating by Street

共享/Sharing

以共享为概念，对传统空间类型进行分类，通过对传统空间的分析转译实现空间的共享，同时在打通的方向上创造空间的流动性，设计出丰富有趣的空间体验。

This scheme is based on the concept of sharing. By means of classifying traditional space categories, this scheme is intended to achieve space sharing on transferring the traditional space based on analyzing, and supplies a bunch of interesting space experiences by increasing the liquidity in the open direction.

景廊/View Corridor

以视线为概念，充分利用长汀老城之中可以看到的所有景色，希望人在其中行走、吃饭的时候可以感受到长汀的特色。

This scheme is based on the concept of sight. By making full use of all the scenes of the old city in Changting, this scheme will allow the people who are walking or eating in the view corridor to feel the features of Changting.

折巷/Meandering Street

以折巷为概念，从长汀老城狭窄而曲折的传统街道空间出发，希望将这种有趣的街道空间延伸到街廊内部，同时充分利用院落，创造有传统韵味的现代酒吧街。

The concept is based on meandering street, starting from the traditional street space of Changting which is narrow and winding, hope to extends the interesting street space into the district, and at the same time make full use of the courtyard, creating a modern bar street with traditional flavor.

沿街美食/Foods alongside the Street

以沿街吃为概念，以街道加餐桌为基本尺寸在场地中进行缭绕，希望在这个可以为旅游提供团餐的场地中创造不同的餐饮体验。

This scheme is based on the concept of foods alongside the street. By placing streets full of dining tables and chairs surrounding the landscape sites, this scheme is intended to provide different dining experiences for tourists groups.

暮色晨光/Day and Night in Fragmented Light

方案以光为概念。在长汀老城调研中发现了多种不同的光进入方式，希望通过对光的空间的归纳变形，提供给人以光的韵律的酒吧体验。

This scheme is based on the concept of lights. A variety of light entering methods were found in the survey about the old city of Changting. This scheme is intended to reflect the summarizing and transformation of lights in space, providing a light rhythm bar experience.

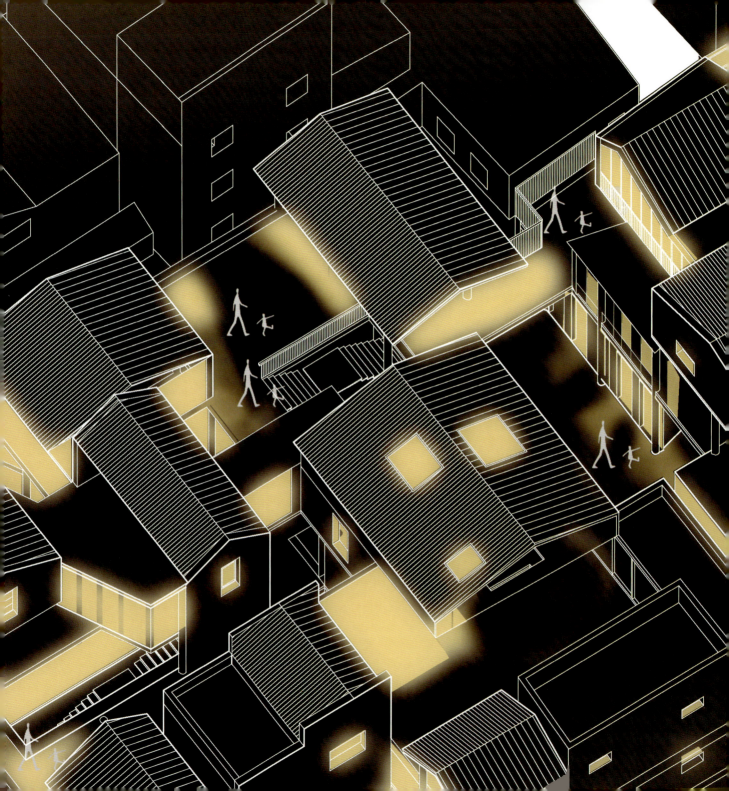

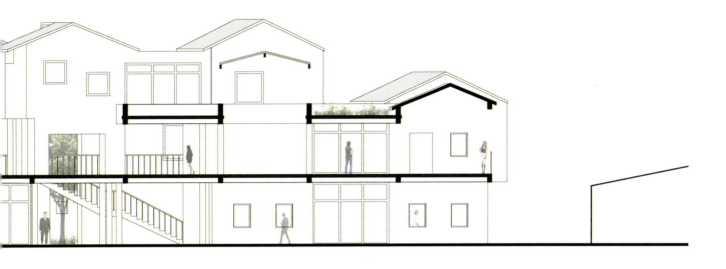

本科毕业设计 GRADUATION PROJECT

数字化设计与建造
DIGITAL DESIGN & BUILDING

吉国华

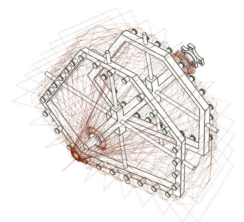

当今，数字化设计与建造相关教学与研究的大规模开展给当代建筑学带来了诸多较有意义的探索，主要体现在两个方面：一方面，它扩大了建筑从设计到建造的创作路径与实现手段，促使两者的互相关联和高度整合；另一方面，它充分显示了建造之于设计所充当的限定与创造的双重角色，能为常规技术、新技术以及新材料等不同制约条件之下的建筑设计带来更多可能性。在这一背景下，2017年南京大学毕业设计教学更加注重建筑数字技术下设计建造自身具有的关联性，强调如何用建造的逻辑表达设计的理念。

自2012年起，南京大学建筑学的本科毕业设计一直在探索数字化设计与建造的教学研究，并开展了一系列专题化的教学实践，包括研究参数化生成结构形态的张拉整体结构与互承结构专题，探讨不同数控建造工具应用的数控机床（CNC）、3D打印与机械臂建造等专题。在此基础上，2017年南京大学本科毕业设计的数字化教学则尝试不再限制某种设计主题或数控建造工具，而是回归到建造这一建筑学的基本命题，强调从数字化的角度认知与建造实践相关的基本知识，让学生在灵活运用新的数字化工具进行设计的基础上，将设计重点放在建造合理性引导的空间形式和建构形式上，从而将建造作为数字化设计建构的一种出发点。

本次课程以"休息亭"为题，要求参与的9位同学自由寻找校园内的一块场地，放置与场所契合的3m×3m×3m的构筑物，并规定每位同学须单独完成相关的阶段作业及最终实物搭建，以体验和完成从数字化设计到数字化建造的全过程。整个课程以讨论为教学手段，以模型为研究媒介，希望在形式生成与建造验证这一往复的过程中，引导学生逐步形成关联设计与建造过程的协同思维模式，以建立寻求物质逻辑合理性的主动思考。

整个教学以建造为核心展开，主要包含前后连接的3个训练环节：（1）从设计到建造层面对案例进行分析与模拟，引导学生体会形体生成与建造逻辑之间的内在关联，提炼案例的建造原型；（2）基于场地调研，学生以1:10比例的模型作为建造验证的媒介，完成从建造原型到数字化表达的设计研究；（3）基于真实材料的实践操作，完成1:1比例的数控加工与实体搭建。在这三个环节中，前一个环节的结果可作为下一个环节的输入，这样有助于引导学生循序渐进地掌握数字化设计与建造之间的转换，并在各个阶段探讨相应的建造问题。

本课程第一阶段的训练核心是通过案例的学习重新审视数字技术，发现数字化设计与建造之间隐藏的关联性。基于以往的教学经验，该阶段直接让学生对所选案例进行综合的分析与学习，因此有别于首先以数字化软件知识讲解的常规授课模式。它不仅要求学生分析案例形体的生成逻辑后在计算机中实现，同时要求学生通过1:10的模型搭建，提炼案例的建造原型，以理解设计与建造之间的关联性。这一阶段的的在于帮助学生初步掌握数字化设计的相关思维方式，重点是引导学生建立设计一造两者之间的互动思维。此外，针对学生并未深入接触过数字化设计及相关工具这背景，该阶段可以帮助学生掌握编程原理、几何工具、算法机制等数字化设计基础理。

本课程第二阶段的训练核心是基于设计与建造关联性的设计研究，它要求学面对实际场地，基于建造原型，把材料、节点、力学逻辑等作为设计的出发点，并虑如场地、功能、空间等典型的建筑设计限定要求，创造出新的数字化方案。这一段中，如何强调与贯彻建造这一核心概念是关键问题。因此，本课程在要求学生形设计的同时，还需要基于1:10比例的模型推敲来完善解决方案。这种研究既包括对料、结构、构造的自身合理性研究，也包括对它们所能产生的设计表现力研究。因在研究过程中，学生既要设计又是建造：在设计的时候，除了进行形体生成的设计，同时需要思考这一特定形式如何进行物质构建，并不断通过模型来验证建合理性；而在模型验证的时候，又要通过材料操作与节点设计，思考这一特定建辑赋予形式表达的可能性，从而对设计进行反馈并完善。

本课程第三阶段的训练核心是实体搭建，基于物质性的操作与体验，进一步加学生对形式生成逻辑的认知、对建造原型的理解以及对构造的精准把控。这一阶段为3个步骤。（1）节点试做，通过1:4比例节点的实体搭建以对建造进行深入研究与进，重点是让学生直面材料，尽管在研究阶段已初步确定材料类型及其节点，但比例由1:10至1:4时，这一改变意味着仍需对材料性能与建造方法进行探索。（2）1比例的整体搭建，目的是让学生以此验证设计的整体建造性能。（3）1:1比例的实搭建，学生通过对材料的机械加工与手工操作，在切身体验中得以解决真实的建造题，这使得设计在建造过程中得以延续，而不再仅仅是一种绘图技能的训练。在这过程中，学生对待材料的加工方式、节点设计、搭建顺序等设计思考均会反映到建形体的外在表达上，并更为直观。

本课程的教学实验显示了建造作为数字化设计教学目标的巨大潜力，它能以更具体、高效的方式让我们直面数字技术引发的设计思维转变，其不仅仅体现于形式成的逻辑方法，更在于建筑学正走向设计与建造的体系整合。本课程同样显示了数建造可以很好地结合对建筑本体问题的思考，而这一思考的结果反过来也促进了建形式的感知与设计能力的培养。与此同时，尽管此次教学设定以建造原型作为设计的起点，试图引导学生将注意力集中在设计与建造之间的关系上，但在对建造原型一一概念相对陌生的情形下，随着形式复杂度的逐渐提高，学生还是更容易陷入对形美学的追求，而忽略了建造方面的问题，从而导致形式逻辑和构造合理两者的不一。因此，对于如何在数字化设计过程中贯彻建造原型的问题仍有待在未来继续研讨改进。

day, the large-scale teaching and research of digital design and building provide ntemporary architecture with many meaningful explorations, which is mainly lected in two aspects: firstly, it expands the creating route and realization means m architectural design to building, promotes mutual relation and highly integration both; secondly, it fully shows the dual role of building to design, which is limitation d creation, and can bring more possibilities to the architectural design under ferent restrictions as common technology, new technology and new material, etc. der this background, the graduation design teaching of Nanjing University in 2017 aches more importance to the relevance of design building under digital technology architecture, and emphasizes the idea that how the building logic expresses sign.

nce 2012, the architecture undergraduate graduation design of Nanjing University s explored the teaching research of digital design and building, and conducted series of special teaching practices, including research on stretch-draw overall ucture and reciprocal structures of parameter generated structure form, discussing different numerical controlled building tools applied CNC, 3D printing and echanical arm building, etc.. On this basis, the digital teaching of undergraduate aduation design of Nanjing University in 2017 tries not to limit some design eme or numerical controlled building tool, but returns to building, the elementary position of architecture, emphasizes on the basic knowledge of cognition and ilding practice from digital perspective, on the basis of making students flexibly ply new digital tool for design, focuses design on building rationality guided space m and structure form, so as to start with building as digital design structure.

this course, "half-way house" is the theme, 9 participating students are required to ely search a site in the campus and place 3m×3m×3m structure that tallies with the e, and every student shall independently complete the operation at relevant stage d the building of final material, so as to experience and complete the whole process m digital design to digital building. By means of discussion, with the model as e research medium, the whole course expects to guide students to gradually form e coordinative thinking mode of related design and building to establish positive nking of seeking for the material logic rationality in the reciprocating process of m generation and building verification.

e whole teaching is conducted with the core of building, mainly including 3 nnected training links: (1) Analyze and simulate the case from design to building, ide students to experience the internal relationship of form generation and building gic, distill building prototype of case. (2) Base on field survey, with 1:10 model as e medium of building verification, complete design research from building prototype digital expression; (3) Base on the practical operation of real material, complete l numerical control processing and entity building. In these 3 links, the result of the evious link can be used as the input of the next link, which is conductive to guide udents to progressively master the conversion between digital design and building, d discuss relevant building issue at different stages.

e training core of the first stage of this course is reviewing digital technology rough case learning, finding the hidden correlation between digital design and ilding. Basing on the past teaching experience, at this stage, students directly make mprehensive analysis and learning of selected case, therefore, it is different from e common teaching mode in which digital software knowledge is taught. It does not ly require students to analyze the generation logic of case form and realize in the

computer, but also require students to distill the building prototype of case through 1:10 model building, so as to understand the correlation between design and building. The purpose of this stage is to help students preliminarily master relevant thinking mode of digital design, focus on guiding students to establish the interactive thinking of design-building. In addition, under the background that students do not deep contact digital design and relevant tool, this stage can help students master programming principle, geometric tool, algorithm mechanism and other digital design basic principle.

The training core of the second stage of this course is the design research based on correlation between design and building, it requires students to face the actual site, base building prototype, take material, node, mechanical logic, etc. as the design start, consider the typical architectural design limit requirement as site, function, space, etc., and create new digital plan. At this stage, the core concept of how to emphasize and implement building is the key issue. Therefore, while requiring form design of students, this course also requires students to deliberate on the basis of 1:10 model to improve the solution. This research includes the research on rationality of material, construction, structure, and also the research on design expression they can produce. Therefore, in the research, students design and build: in the design, besides digital design of form generation, they should also think the way of material building of this specific form, and verify the rationality of building constantly through model; in model verification, they should think the possibility that this specific building logic endows form expression through material operation and node design, so as to feed back and improve the design.

The training core of the third stage of this course is solid building based on material operation and experience, which further deepens the student's cognition of form generation logic, understanding of building prototype and precise control of structure. This stage is divided into 3 steps. (1) Node trial production, in-depth research and improvement of building through 1:4 node building, focus on making students face the material, despite the material type and node have been preliminarily confirmed at the design research stage, when the scale is turned from 1:10 to 1:4, it means that exploration on material performance and building method is still required. (2) Overall building at scale of 1:4, with the purpose of making students verify the overall building property of the design. (3) 1:1 actual building, students solve real building problem in personal experience through machining and manual operation of material, which continues design in the building process instead of only training on drawing skill. In this process, the design thinking of students on material processing mode, node design, building sequence, etc. will be reflected on the external expression of architecture form, which is more perceptual.

The teaching experiment of this course shows the huge potential of building as the target of digital design teaching, it can make us directly face the change of design thinking caused by digital technology in a more specific and efficient method, it is not only reflected in the logic method of form generation, but also in the aspect that architecture is leading to the system integration of design and building. This course also shows that digital building can well combine the thinking of architecture ontology problem, and the result of this thinking also promotes the cultivation of perception and design ability of architectural form. At the same time, despite the teaching sets building prototype as the start of design research, tries to guide students to focus on the relationship between design and building, in the circumstance that the concept of building prototype is relatively unfamiliar, with the gradually increased complexity of form, students are easy to be caught in the pursuit of form aesthetics and ignore the problem of building, which leads to the inconsistence of form logic and structure rationality. Therefore, how to implement building prototype in the digital design is still to be discussed and improved in the future.

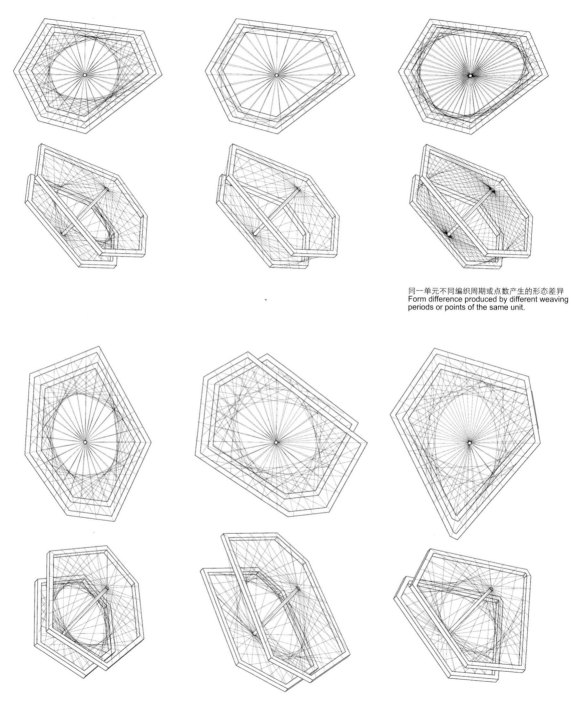

同一单元不同编织周期或点数产生的形态差异
Form difference produced by different weaving periods or points of the same unit.

同样点数和周期不同形态单元编织。
Unit weaving of the same points and period and different forms.

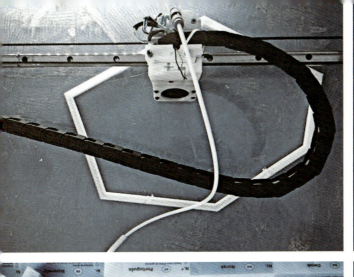
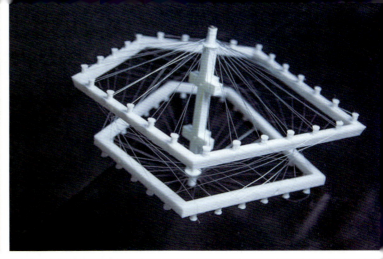
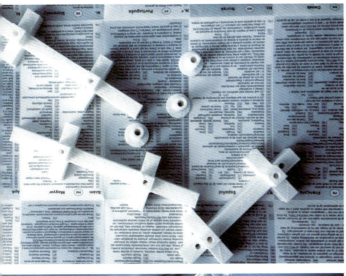
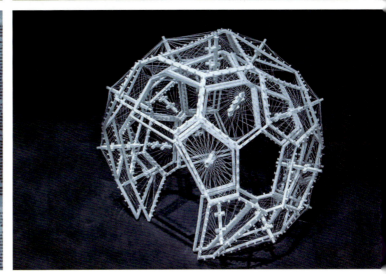
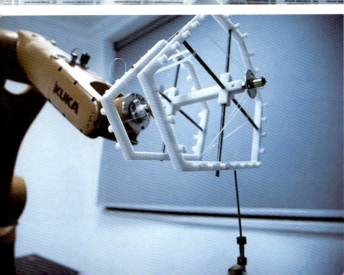
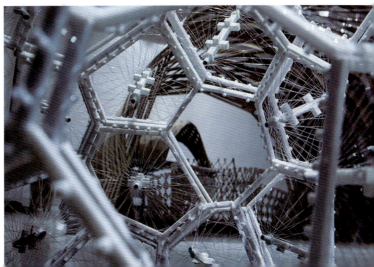

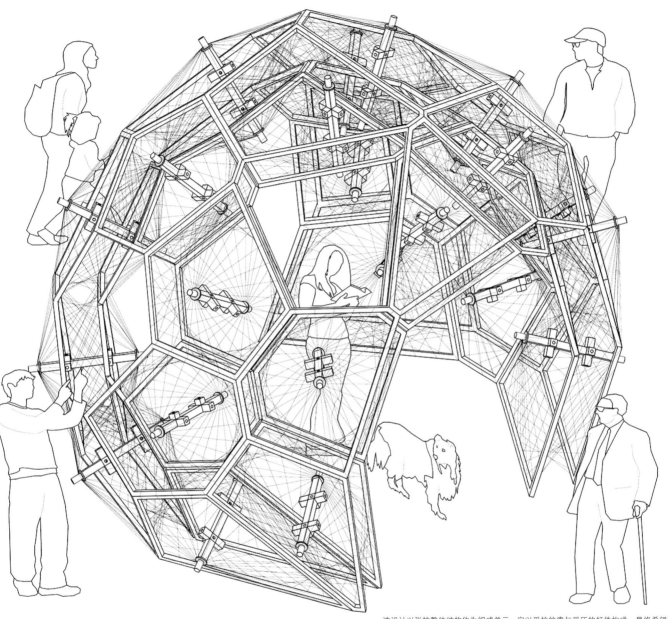

该设计以张拉整体结构作为组成单元，它以受拉的索与受压的杆件构成，最终希望通过机械臂编织受拉索而达到单元与整体的结构稳定。
This design uses overall structure of tension as the component unit. It consists of pulled rope and stressed pole, and finally realizes stable structure of unit and whole through mechanical arm weaving pulled rope.

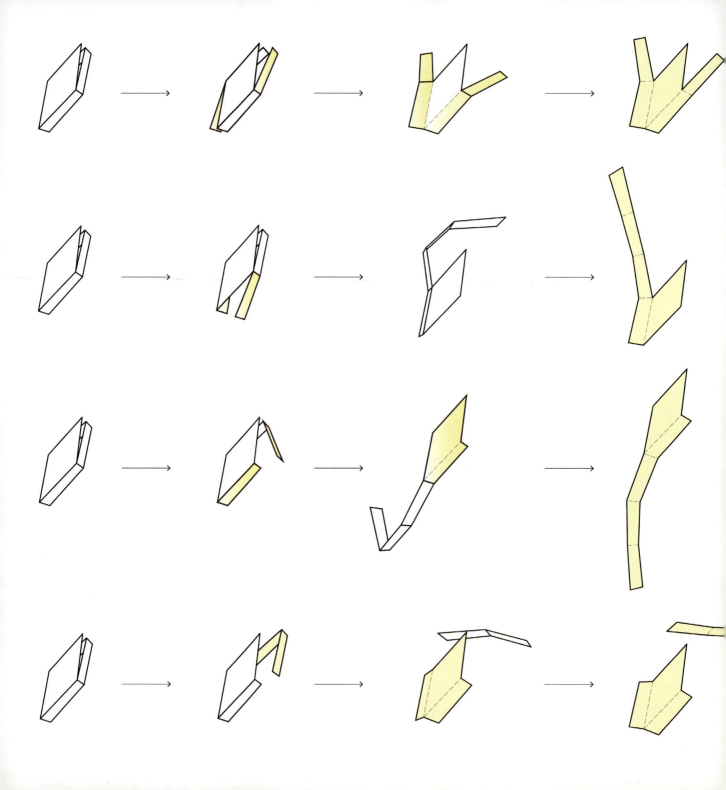

该设计通过树木的相对位置建立参数化关系,并利用Kangraoo软件生成符合受力要求的壳体形式。
This design establishes parameter relation through relative position of trees, utilizes Kangraoo software to generate shell form that meets the stress requirement.

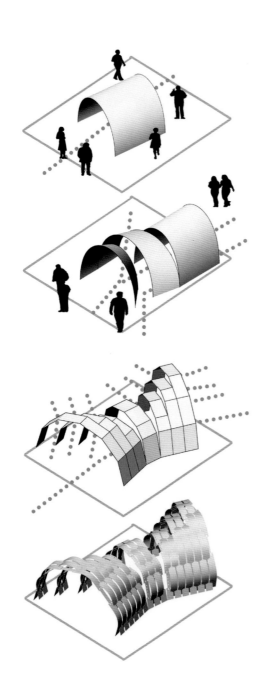
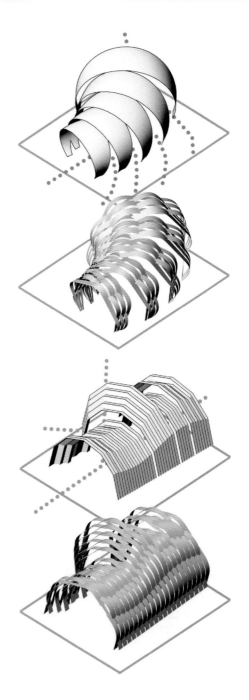

该设计利用弹性线方程,对构筑物形体进行了计算性生成,从而使弹性材料的预制弯曲成为可能。
This design utilizes elastic line equation, calculates and generates the structure form, so as to make prefabricated bending of elastic material possible.

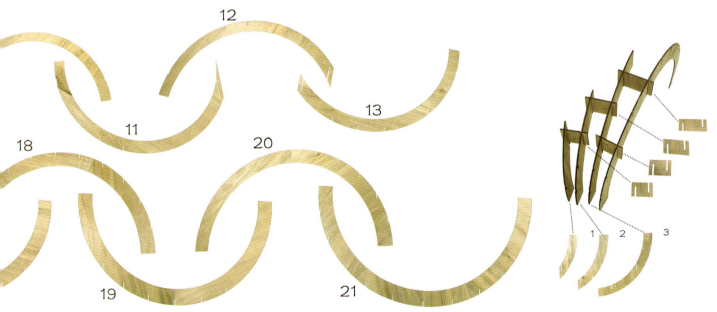

建筑设计研究（一） DESIGN STUDIO 1

记忆·场所·叙事
MEMORY · PLACE · NARRATIVE

鲁安东

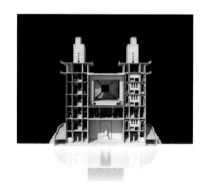

在我们这个时代，建筑学不应再是铺陈记忆的艺术，而是一种提醒的艺术。它为记忆的显影提供参照系，帮助人与场所之间建立起深沉的情感归属，进而得以触及人的灵魂。它是一种叙事的艺术。

记忆既是对当代建筑学的巨大挑战，也是其物质建造的终极命题。我们需要新的形式、叙事的形式，来涉足未知的记忆领域。

从记忆的角度来说，物质、活动、感动与存在之间不存在清晰的边界划分。它们共同构成了一次浮现，通过现在的再想象连接起过去与未来。

记忆的建筑学关注记忆在场所中的发生机制，而不是对记忆内容的选择性再现。记忆的建筑学关注主体参与的精神体验，而不是映射着环境的感知体验。它需要超越物质、空间与符号的新建筑语言，来沉浸、互动、质询与映射主体。正如现代主义背离了语言，用感知替代了人与场所之间的意义关联，记忆的建筑学将重新回到语言的领域，并召唤叙事的心灵。

场所记忆的意义建构应该是开放的，而不是被给予的。设计和技术，应该帮助人找回自我与世界的归属感。它们必须走向存在的维度。这是当代建筑学的使命。

此次"概念设计"课程选择了南京长江大桥作为研究对象。对南京长江大桥的记忆交织着宏大的历史记忆与亲密的个体记忆，曾是许许多多人建设、学习、旅行、工作、生活乃至想象的一部分。本课程从大桥记忆入手，探讨空间记忆的特征、发生机制和意义建构，最终设计一个当代的、公共的和富有创造性的记忆场所。

课程介绍：
为期八周的教学分为四个阶段：
前期阶段，案例研究（1周）：结合场地调研，寻找并分析案例，对记忆场所提出个人理解。
第一阶段，独立研究（2周）：提出初步概念，结合大桥场地对概念进行测试。
第二阶段，方案初期（2周）：依据概念相关性，将同学分为两人一组，并发展出新概念。
第三阶段，方案中期（2周）：结合场地，通过示范性设计探索概念的潜力。
最终成图（1周）：每组通过一套十张图纸，对概念进行系统的表达。

At this time, architecture should not be an art to display memory, but an art reminding. It provides reference to manifest memory, helps people and place establ profound emotional belonging, so as to touch the soul. It is an art of narration.

Memory is a huge challenge to contemporary architecture, and also the ultima proposition of its material building. We need new form, form of narration to step in unknown field of memory.

From the perspective of memory, substance, activity, touch and existence have clear boundary. They compose an appearance, and connect the past and the futu through current re-imagination.

Architecture of memory concerns the occurrence mechanism of memory in t place instead of selective reproduction of memory content. Architecture of memo concerns the spiritual experience of entity participation instead of reflecting t emotional experience of environment. It needs new architectural language t surpasses substance, space and symbol to immerse, interact, question and map entity. Just as modernism deviated from language, perception replaces the mean relation between people and place, architecture of memory will return to the field language and recall the soul of narration.

The meaning construction of place memory should be open instead of being giv Design and technology should help people retrieve the sense of belonging of s and world. They must walk to the dimension of existence. This is the mission contemporary architecture.

In the course of conceptual design, Nanjing Yangtze River Bridge is selected as t research object. The memory of Nanjing Yangtze River Bridge interweaves gra history memory and close personal memory, and it is a part of construction, learni travel, work, life and even imagination of many people. This course starts with t bridge memory, discusses characteristics, occurrence mechanism and meani construction of space memory, and finally designs a contemporary, public a creative memory place.

Course introduction:
Eight-week teaching is divided into four stages:
Preparation stage, case study (1 week): combine with field survey, search and analyse cases, propose personal understanding of memory place.

First stage, independent research (2 weeks): propose preliminary concept, test the concept according to the bridge place.

Second stage, early period of plan (2 weeks): according to the concept correlation, two classmates in a group, develop new concept.

Third stage, mid-term of plan (2 weeks): combine with the place, explore the potential of concept through demonstrative design.

Final drawing output (1 week): each group systematically expresses concept through ten drawings in a set.

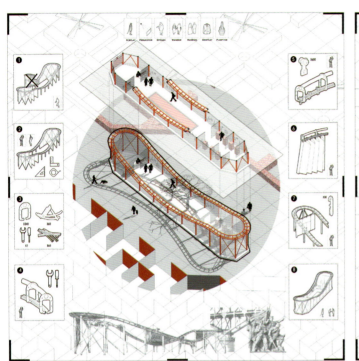
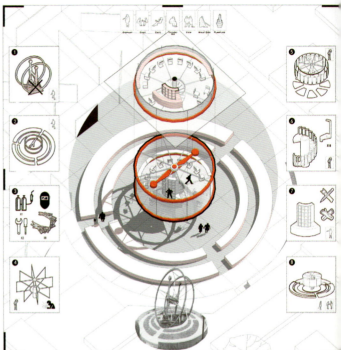
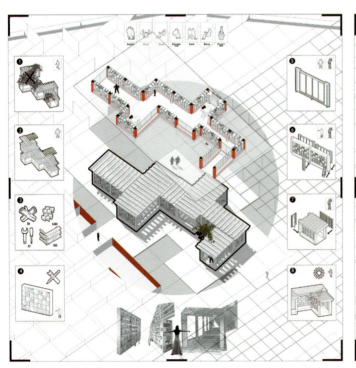

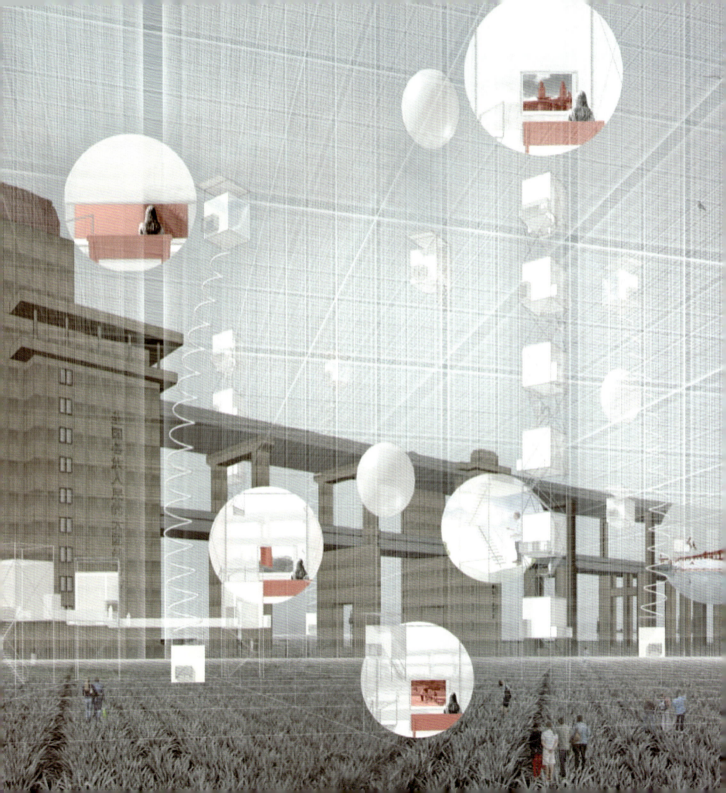

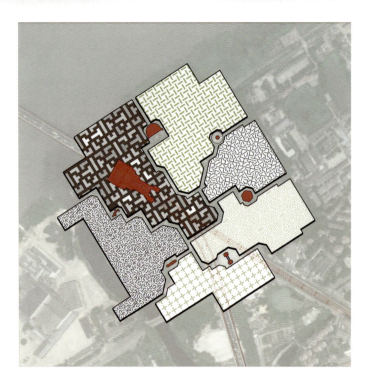
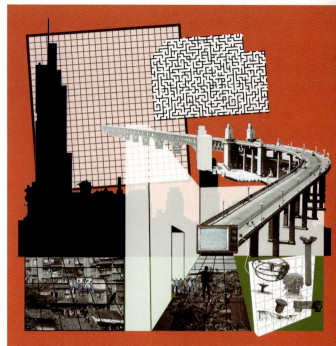

建筑设计研究（二） DESIGN STUDIO 2

建构研究"低技建造"设计研究
DESIGN RESEARCH ON "LOW-TECH CONSTRUCTION" IN CONSTRUCTION RESEARCH

傅筱 孟宪川

2015年，在周凌、赵辰两位老师的努力下，课程组得到了相关资金和场地的支持，"低技建造"课程选择在浙江莫干山南路乡"60亩农田服务设施规划"场地内，以竹结构为主实地建造"山野乐园"景观小品，供游客和儿童使用。2016年，在鲁安东老师和南京大学建筑设计研究院程向阳总建筑师的引荐下，课程组得到溧水区住房和城乡建设局和溧水城建集团的支持，在江苏溧水无想山某场地内建造以竹结构休息亭为主的景观小品，供游客使用，并在无想山山顶设计一个木构瞭望台，供游客休息眺望。另一组"低技建造"建造课题组由陈浩如老师带领，在浙江临安太阳公社建造竹构游客餐厅。目前，两个课题组已经完成相应的技术图纸，但由于高温天气等原因，均未进入实际建造阶段。通过两年的教学和实践，我们将"低技建造"设计课程的些许体会在此小记，供读者参考。

所谓"低技建造"设计课程，其目的是让学生通过亲身建造，真切体会到真实材料和真实尺度的意义，并建立起建造是衡量设计的核心标准的认知。之所以强调"低技术"，并非出于学术意义的考量，而是便于让学生能够亲身参与建造，通过身体与材料的直接接触，培养对材质、尺度、重量等性能的切身体会。因此，要符合低技建造的教学条件有以下几点：

（1）建造规模不宜太大，一般以简易休息亭的规模为宜，建造时尽量不借助机械设备就能够完成；

（2）建筑功能必须简单且明确，因为本课程训练的目标是建造技术而非功能布局，明确功能定位有利于学生将精力集中在建造与材料、结构、构造的关系上；

（3）材料必须易于学生手工操作，从材料自重选择到节点加工都尽量在学生的体能范围之内；

（4）建造的周期不宜过长，一般控制在半个月以内为宜。如果建造时间太长将让学生身心疲惫，因为学生毕竟不是专业的技术工人，并不具备长时间操作的职业技能。

能够符合上述条件的建造体系并不多，木构、竹构将是较为理想的建造体系，砌砖虽然看似易于操作，但是如果要让学生砌筑形成一个有覆盖的符合人体尺度的结构体系，其难度不言而喻，如果仅仅砌筑一些片断墙体，对学生认知结构体系、构造做法的训练是不够的。木构的结构逻辑容易控制，有科学合理的技术规程，构造方式比较成熟，节点种类比较丰富，从整体结构体系到节点的力学连接都容易让学生理解，从而让学生建立材料与受力、力流传递与节点受力的关联思考。木构略显不足的是构件的加工需当地工厂配合，学生参与的程度会有所降低。如果建造地点在本地充分利用高校木工实验室让学生自己加工构件将是比较理想的选择。此外，木构造价偏高也是需考虑的因素之一。

与木构相比，竹构的优势是就地取材，造价低廉，对建造精度要求不高，易于场手工操作，这也是近几年流行竹构建造教学的原因之一。然而，竹构的优势恰恰是它的劣势。首先，每根原竹都有所不同，力学性能难以准确描述，所以原竹结构系是以经验为技术支撑，无法形成技术规程；其次，同一根竹子分大小头，而且竹是一种很不稳定的材料，容易变形开裂，所以原竹建筑的构造节点是难以去"设计"的，这也是老百姓都用绳子绑扎竹子的原因所在，绳子非常容易适应竹子材料的不则形，并且廉价、易于操作以及修补更换，百姓的智慧不容小觑。

鉴于竹构的上述两个特点，课程组制定了两个教学要点：第一，放弃对竹构节点的追求，尽量避免为节点而节点的设计情结，尊重材料最合适的做法以及最合适学搭建的做法。为此，我们建议学生以绑扎和长螺杆两种连接方式为主，辅助以传统竹销连接，并注重竹构体系与大地的连接关系。第二，为了便于学生理解结构体系力流传递的关系，课题组引入了Karamba结构分析软件，该软件可以在设计过程实时模拟出结构体系的受力分布，包括竹子的性能也可以进行模拟，虽然不是十分确，但是对于学生理解受力问题确实起到较大的帮助。

研究生来自于全国各个高校，学生的基础并不完全相同，在教学过程中，教师能为了教学成果而急于推销自己的设计认知，而应通过启发式教育，放手让学生去考，当某个学生的设计率先达到教学预期，立刻抓住其要点进行点评和讨论，这样能让学生深刻理解相关知识。"低技建造"课程主要解决了两个问题，从设计思维度改变了学生用大脑中已有的造型去思考问题的习惯，逐渐学会从场地、结构、构以及建造综合推导设计的方法；从实地建造的角度让学生体会到材料和尺度不是抽的专业术语，而是对自然规律和体验最真实的描述。

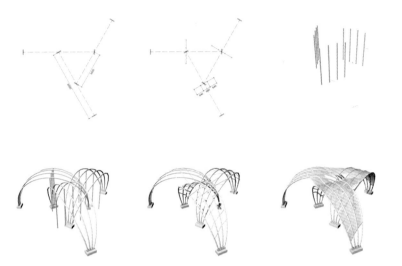

施工步骤：
1. 放线定位，浇筑基础
2. 定位临时支架，固定主体结构角度
3. 搭建临时支架
4. 搭建主体结构，固定角度
5. 搭建次级结构，撤掉临时支架
6. 编织覆盖面

Construction steps:
1. Setting-out and positioning, foundation casting
2. Position temporary support, fix angle of main structure
3. Build temporary support
4. Build main structure, fix angle
5. Build secondary structure, remove temporary support
6. Weave cover

2015, under the efforts of teacher ZHOU Ling, ZHAO Chen, the course team obtained relevant funds and site support, the site of "low-technology building" course was selected to "60 mu farmland service facility planning" in Nanlu Village, Mogan Mountain, Zhejiang, "mountain and plain garden" featured landscape was built in bamboo structure for tourists and children. In 2016, introduced by teacher LU Andong and chief architecture CHENG Xiangyang of Architectural Design Research Institute of Nanjing University, the course team was supported by Lishui Housing and Urban-Rural Development Bureau and Lishui Urban Construction Group to build featured landscape subject to bamboo half-way house on some site in Wuxiang Mountain, Lishui, Jiangsu for the tourists, and design a wooden observatory on the top of Wuxiang Mountain for tourists to rest and overlook. Another group of "low-technology building" task team was led by teacher CHEN Haoru to build bamboo tourist restaurant in Sun Commune in Lin'an, Zhejiang. At present, both task teams have completed relevant technical drawing, but the actual building is not started yet due to high temperature. Through two years of teaching and practice, we will share some experience in "low-technology building" design course here for the reader's reference.
The purpose of so-called "low-technology building" design course is to make student personally feel the meaning of real material and real dimension through building by themselves, and establish the concept that building is the core standard to weigh design. We emphasize "low technology" to make students personally participate in the building and cultivate their experience in material, dimension, weight, etc. through direct body contacts with the material instead of considering the academic meaning. Therefore, the teaching conditions that meet low-technology building include:
(1) The building scale should not be too large, generally should be the scale of simple half-way house, and the building can be manually completed instead of mechanical equipment;
(2) Building function must be simple and clear, for the purpose of the course training is building technology instead of functional layout, clear function is favorable for students to focus on the relationship of building with material, construction, structure;
(3) Material must be easy to manually operate for students, and shall be within the scope of physical ability of students from material selection to node processing;
(4) The building period should not be too long, generally should be controlled within half a month. If the building period is too long, students will be tired, for students are not professional technical worker and have no long-term operation skill.
There are not many building systems that can meet aforesaid conditions, wooden structure, bamboo structure are ideal building system, it seems that brick building is easy to operate, but it is difficult to make students build and form a covered structure system that meets human dimension. Only building some segment wall is insufficient to train students on the cognition building system, structure method. The structural logic of wooden structure is easy to control, there is scientific and rational technical specification, the structure method is mature, the node is rich, the mechanical connection from overall structure system to node is easy to understand, so that students can establish the correlative thinking of material and stress, force flow transfer and node stress. The inadequacy of wooden structure is that the processing of wooden structure component requires the cooperation of local factory, and the degree of participation of student will be reduced. If the building site is local, fully utilizing the woodworking laboratory of the university to make students process component themselves will be an ideal choice. In addition, higher cost of wooden structure is one of the considerations.
Compared to wooden structure, bamboo structure can use local material at low cost, the requirement on building precision is not high, and it is easy to manually operate on the site, this is also one of the reasons that bamboo structure building teaching is popular in recent years. However, the advantage of bamboo structure is also its disadvantage. Firstly, each piece of raw bamboo is different, it is hard to precisely

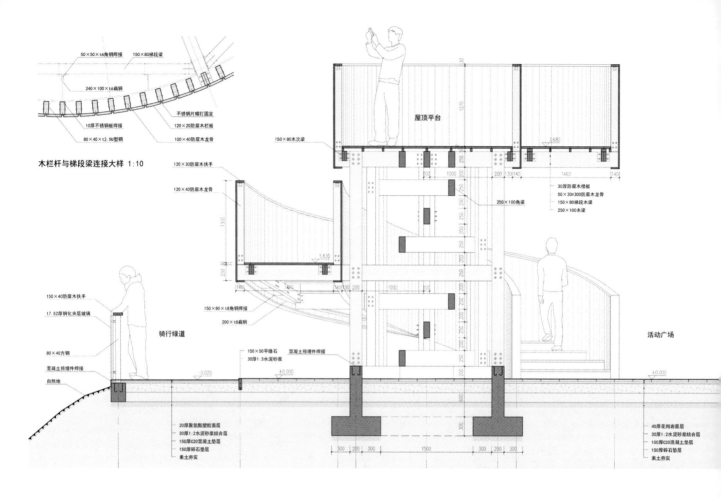

describe the mechanical property, therefore, experience is the technical support of raw bamboo structure system, and technical specification cannot be formed; secondly, the same piece of bamboo has large end and small end, while bamboo is an unstable material, easy to be deformed and cracked, therefore, it is hard to "design" the structure node of raw bamboo architecture, so common people use rope to bind bamboo for rope is easy to adapt to the irregular shape of bamboo material, it is cheap and easy to operate, repair and replace, the wisdom of common people cannot be underestimated.

In view of aforesaid two characteristics of bamboo structure, the course team develops two teaching essentials: firstly, give up on pursuing bamboo structure node, try to avoid the design complex of node, respect the most suitable method for material and the most suitable building method for student. Therefore, we suggest students using binding and long bolt, coupled with traditional bamboo pin, and pay attention to the connection between bamboo structure system and the earth. Secondly, to facilitate students to understand the relationship between structure system and force flow transfer, the course team introduces Karamba structure analysis software, which can realize real-time simulation of stress distribution of structure system in the design, including simulation of bamboo property, not very precise although, it is a great he for students understanding the stress.

Postgraduates are from different universities and have different foundations, the teaching, teachers cannot be anxious to promote their design cognition for th teaching result, instead, teachers should make students think through enlightenme education, when the design rate of some student reaches the teaching expectatio teachers should immediately grasp the essential to comment and discuss, so as make students deeply understand the relevant knowledge. "Low-technology buildin course mainly solves two problems, changing the habit of students to use the existir model in brain to think from the perspective of design thinking, making studen gradually learn to comprehensively infer and design from site, construction, structur building; making students learn that material and dimension are not abstractiv professional terms but the most real description of natural law and experience fro the perspective of field building.

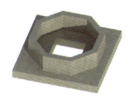

1. 浇筑钢筋混凝土基础
1. Cast reinforced concrete foundation

2. 架设一品梁架
2. Set one beam frame

3. 旋转生成四品梁架
3. Spin and generate four beam frames

4. 连接顶部木梁
4. Connect top wood beam

5. 在主梁间连接段梁
5. Connect section beam between main beams

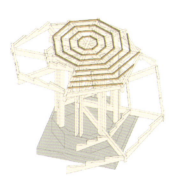

6. 在木梁顶部铺设木龙骨
6. Lay wood keel on the top of wood beam

7. 在梯段梁上铺设木楼板
7. Lay wood slab on the stair beam

8. 在楼梯两侧连接木龙骨
8. Connect wood keel on both sides of stair

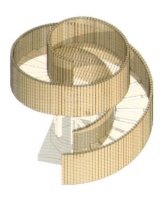

9. 在木龙骨外侧安装木栏板
9. Install wood fence outside the wood keel

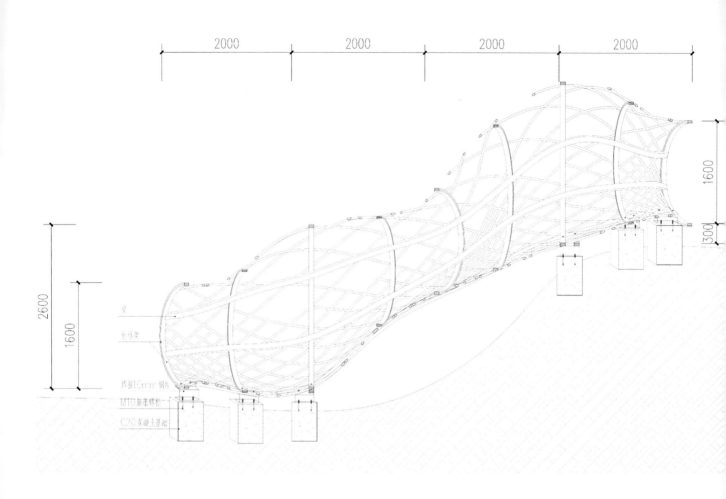
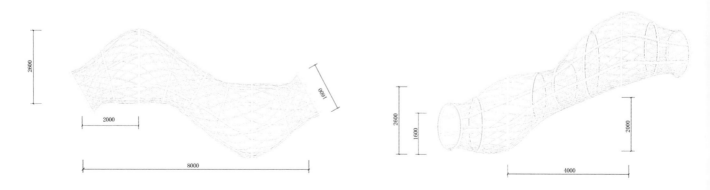

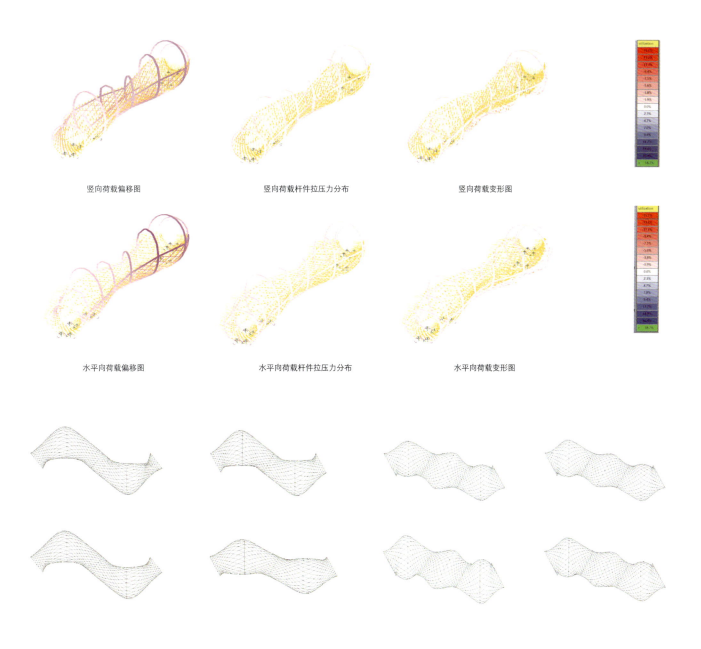

竖向荷载偏移图	竖向荷载杆件拉压力分布	竖向荷载变形图
水平向荷载偏移图	水平向荷载杆件拉压力分布	水平向荷载变形图

设计引入 Karamba 结构分析软件，在设计过程中实时模拟出结构体系的受力分布，包括竹子的性能也可以进行模拟。
The course team introduces Karamba structure analysis software, which can realize real-time simulation of stress distribution of structure system in the design, including simulation of bamboo property.

木构的结构逻辑容易控制,有科学合理的技术规程,构造方式比较成熟,节点种类比较丰富,从整体结构体系到节点的力学连接都容易让学生理解,从而让学生建立材料与受力、力流传递与节点受力的关联思考。
The structural logic of wooden structure is easy to control, there is scientific and rational technical specification, the structure method is mature, the node is rich, and the mechanical connection from overall structure system to node is easy to understand, so that students can establish the correlative thinking of material and stress, force flow transfer and node stress.

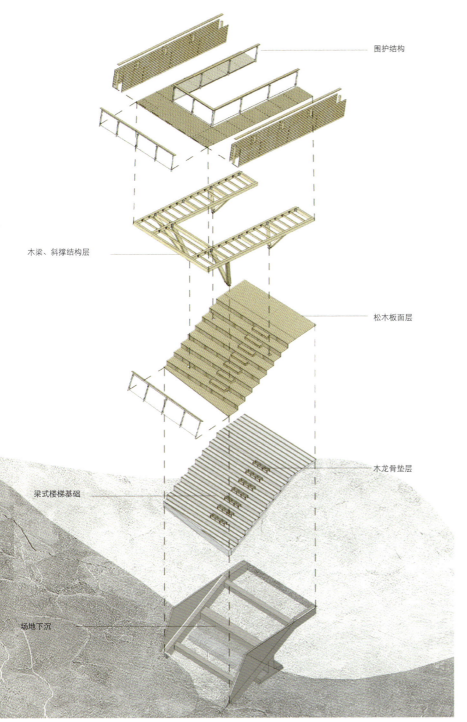

围护结构

木梁、斜撑结构层

松木板面层

木龙骨垫层

梁式楼梯基础

场地下沉

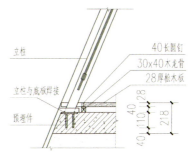

底层栏杆节点大样 Detail of bottom layer rail node

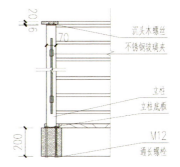

上层栏杆节点大样 Detail of upper layer rail node

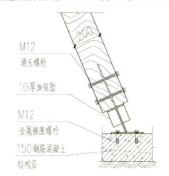

斜撑纵向节点大样 Detail of diagonal braced longitudinal node

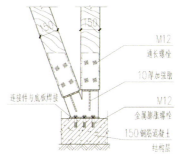

斜撑横向节点大样 Detail of diagonal braced transversal node

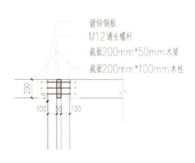
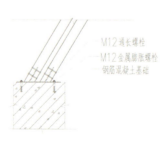

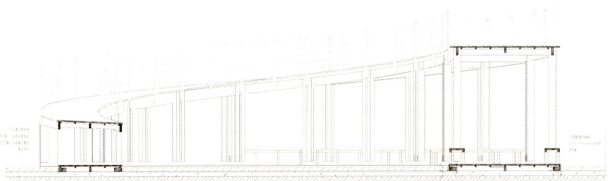

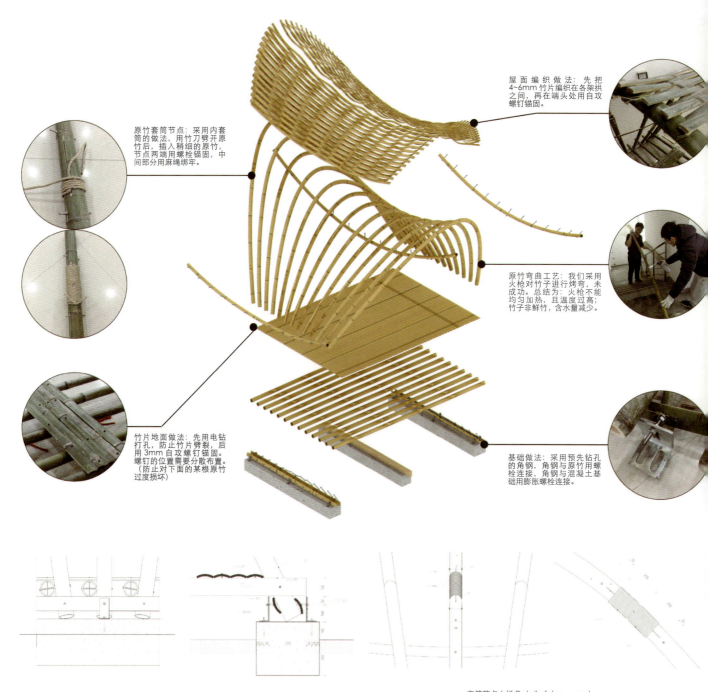

基础部件节点大样 Detail of foundation component node　　套筒节点大样 Detail of sleeve node

研究生国际教学交流计划
THE INTERNATIONAL POSTGRADUATE TEACHING PROGRAM

南京大学建筑与城市规划学院一直致力于最前沿的学术理论研究,并以之解决实际问题。在研究成果输出的基础之上,学院积极参与国际学术交流并致力于为未来培养建筑和城市规划的高水平人才。为了拓展学院的国际学术交流网络并且建立长期的合作交流体系,2016年南京大学建筑与城市规划学院启动了"研究生国际教学交流计划"(IPTP)。IPTP是一个灵活的邀请教学计划,每年都会有10位专家教授接受邀请前来教学访问。

今年开始,"研究生国际教学交流计划"(IPTP)正式实施。已经完成的课程有:

1. 地中海的重生——"绿色街区"
   胡安·高尼、傅筱、徐一品
2. 针对南京城市问题的类型学设计研究
   斯蒂法诺·科尔博、王丹丹
3. 参数化图解静力学
   科朗坦·菲韦、孟宪川
4. 环境交互设计
   若泽·阿尔莫多瓦、唐莲、尤伟
5. 繁荣的后合理化
   费洛里·科萨克、窦平平
6. 城市形态学:城市景观分析法
   谷凯、胡友培

Since its foundation, the School of Architecture and Urban Planning, Nanjing University is committed to cutting-edge academic theory courses, in order to address contemporary issues. Based upon research outputs, the School is actively engaged in international academic exchanges and targets at nurturing high-level professionals in architecture and urban planning for the future. In order to further extend international academic exchange network and to establish a long-term cooperation and exchange mechanism, the School has launched the International Postgraduate Teaching Program (IPTP). The IPTP is a flexible guest-teaching program that includes 10 visiting positions annually.

The IPTP has been carried out since this year. Programs finished include:

1. The Regeneration of the Mediterranean -- "Green Block"
   Juan GOÑI, FU Xiao, XU Yipin
2. Hybrid Typologies for Nanjing City Problem
   Stefano CORBO, WANG Dandan
3. Parametric Graphic Statics Workshop
   Corentin FIVET, MENG Xianchuan
4. Environmental Interactive Design
   José AIMODóVAR, TANG Lian, YOU Wei
5. Re-appropriate the Boom
   Florian KOSSAK, DOU Pingping
6. Urban Morphology: An Analytical Approach to the Urban Landscape
   GU Kai, HU Youpei

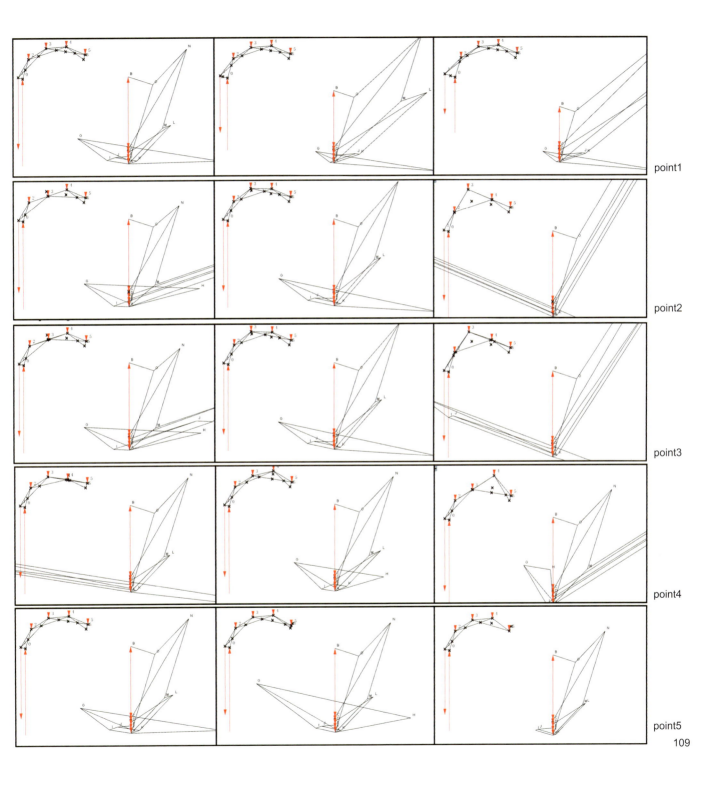

研究生国际教学交流计划课程 IPTP

地中海的重生——"绿色街区"
THE REGENERATION OF THE MEDITERRANEAN —— "GREEN BLOCK"

胡安·高尼 傅筱 徐一品

在工业革命之后，人类科技文明以前所未有的速度发展，幢幢高楼拔地而起，座座城市快速扩张。然而众所周知，因为人类活动的强大影响，我们生活的自然环境正在面临不断恶化的危机。哪怕是在风景宜人的地中海沿岸，那里的城市也不得不面对气候变化带来的挑战。

在这样的情况之下，西班牙的马拉加城市议会联合了巴塞罗那、巴伦西亚、塞维利亚、罗马、都灵等11座地中海城市，率先发起了"地中海都市革新"计划。该计划在每一座城市中设立一个名为"绿色街区"的城市更新试点地块，通过对试点地块的设计研究，将相关成果推广到其他地中海城市中去，一方面保持地中海城市特色，另一方面逐步解决其环境问题。

本次工作营中，胡安·高尼教授带领大家为马拉加的"绿色街区"进行设计和策略探究，从而提高大家对可持续街区的认知，并提高处理地区文脉与绿色节能技术结合问题的能力。本次工作营的目标基地就是位于马拉加的"绿色街区"地块，目前被一些工厂和公司使用，但在规划之中，这里将会成为一个居住社区。该区域规划中占地94 000 m²，并提供963间拥有齐全公共设施配套的住房。它除了在CAT-MED方案中指导系统的价值之外，也将在数量上超过规定的最低要求，达到最佳环境、经济和社会可持续性的等级。

对于参加工作营的同学而言，每一幢楼的位置和相关限制都已经确定，他们需要做的是根据要求对每一幢楼进行细化设计，并且对场地进行详细规划。相关设计应当在有关能源效率、垃圾管理、水资源消耗以及二氧化碳排放等要求的控制之下进行，从而使得城市代谢达到平衡。

After the Industrial Revolution, human's technology and civilization are developing at a unprecedented speed. Cities keep expanding while more and more buildings are under construction. As we know, however, because of the effects of human's activities, the environment we are living in is facing the risk of becoming worse. Even for the cities around Mediterranean coast, they have to face the challenges caus by climate changes.

In that case, Malaga city council manages "Changing Mediterranean metropolis gradually" programme with 11 Mediterranean cities such as Barcelona, Valenc Sevilla, Rome, Turin. The programme sets "Green Block" as a pilot project of urb regeneration in each city and promotes the relevant results to other Mediterrane cities through the design and research of the pilot project. It will help solve environmental problems gradually while keeping the characters of Mediterrane cities.

In this workshop, Prof. Juan GOÑI leads the students to make design and strateg for Malaga "Green Block" area and the participants can learn more about sustaina blocks and improve abilities about solving problems of context and energy-sav technology. The "Green Block" area located in Malaga is in industrial uses a occupied by some factories and a company for now. But in the plan, this area v become a residential community. The Malaga Green Block comprises an area 94,000 m² and includes the intended creation of 963 dwellings with services a infrastructure which, in addition to the values set out in the CAT-MED programm system of indicators, will quantitatively exceed the legal minimums and qualitativ meet optimal environmental, economic and social sustainability levels.

For each participant in this workshop, the buildings' location and related restricts a already set. What they need to do is to make detail design about every building a the area. And the design must be governed by criteria concerning energy efficier and waste management, a reduction in the consumption of water resources and C emissions so that the urban metabolism achieves a balance.

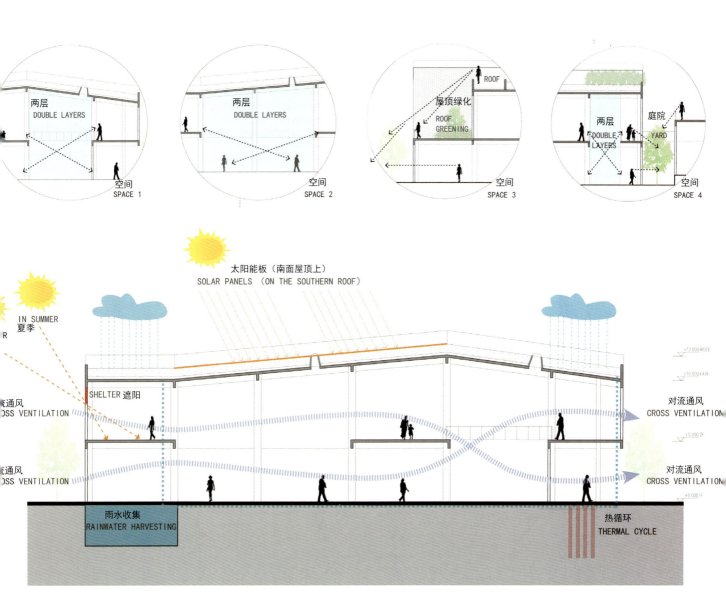

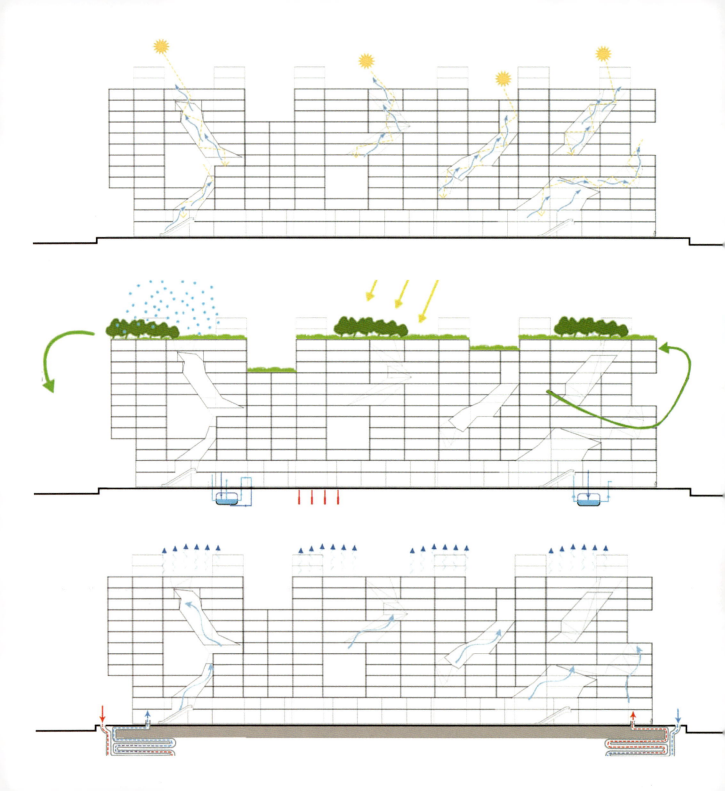

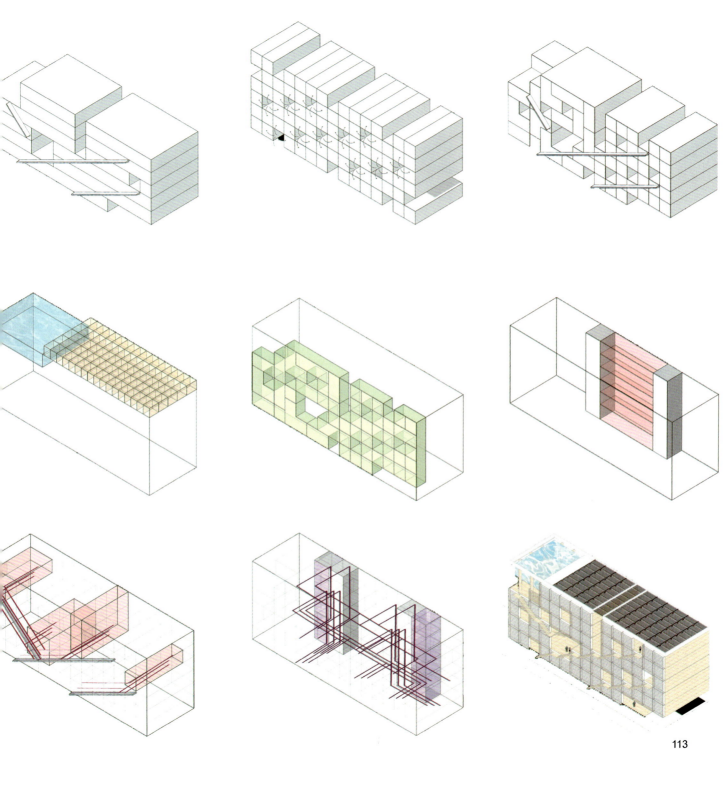

研究生国际教学交流计划课程 IPTP

针对南京城市问题的类型学设计研究
HYBRID TYPOLOGIES FOR NANJING CITY PROBLEM

斯蒂法诺·科尔博　王丹丹

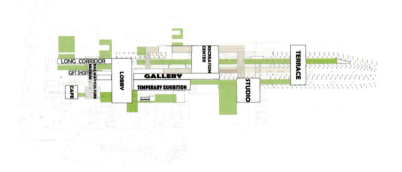

　　基于对当今建筑学发展趋势的思考，斯蒂法诺提出四种空间类型，限定学生选取其中一种，对其加以调整和优化，来解决特定的南京城市空间问题。在10天的密集工作和研讨中，学生们在斯蒂法诺的指导下探索如何用特定的空间类型优化周围城市空间，最终取得优秀的设计成果。
　　四种空间类型分别为：垂直公园、地景建筑、室内景观和可适应场域，而每一种类型都重新诠释了一组传统空间类型。垂直公园打破了"自然与人工"的界限；地景建筑打破了"建筑与城市"的界限；室内景观打破了"室内与室外"的界限；"可适应场域"则用于处理"旧与新"之间的关系。事实上，斯蒂法诺老师从城市肌理中提取的这些空间类型基于他对当今建筑学理论思考的形式表达——即建筑语汇正在从封闭、抽象、静态变得更为开放、流动和充满变化。
　　斯蒂法诺老师和学生们一起讨论，启发学生思考南京潜在的城市问题。学生将先确定要解决的城市问题，再选取相应的空间类型作为解决方案的基础。斯蒂法诺老师强调，对所选类型的思考应贯穿整个设计过程，并与每一个推进的步骤紧密相连。最终，学生选取特定的空间类型来解决不同的城市问题。"河景"方案场地位于秦淮河，该地块为城市高密度住宅区，缺乏医疗服务、文化设施和公共活动空间。为解决这些问题，方案采取"地景建筑"的策略，将河边一处停车场改造为与场地地形融为一体的综合社区中心，提供社区服务的同时使河岸景观与城市形成互动。"线性乌托邦"则选择"室内景观"作为设计策略，通过有机结合地块内已有的各式各样的公共空间，生成动态交互的景观，创造一个线性连续的多功能公共空间，以解决南京大学鼓楼校区宿舍区缺乏公共空间和私人空间等问题。

Based on his thought of the current trend of the architectural discipline, Mr. Stefano provided students with four kinds of spatial typologies, from which students were instructed to choose just one, adjust and apply it to a specific urban space problem in Nanjing. In this ten-day workshop, with the help of Mr. Stefano through intense lectures and tutorials, students managed to explore how to improve a surrounding urban space by means of exploring a kind of spatial typology, and finally produced excellent design work.

Mr. Stefano provided students with four kinds of spatial typologies: vertical park, landform building, interior landscape, and adaptive field. All typologies are mea to re-interpret a group of traditional spatial categories — for the vertical park, breaks the boundary between the natural and the artificial; for the landform building architecture and city; for the interior landscape, interior and exterior; for the adapti field, the old and the new. In fact, those typologies Stefano abstracts from the urb context, are somehow the formal expression of the contemporary theoretical deba That said, architecture is nowadays a open, flexible and soft field, rather than closed, abstract and static object.

Mr. Stefano joined the discussion with students, inspiring them to reflect on the urb problems lurking in Nanjing. At first, students would locate one urban problem, a then select a relevant typology as the premise of solving the problem. Mr. Stefa stated that, throughout the whole design process, the chosen typology should closely connected to each phase of developing the project. Students chose differe types to deal with different problems. The project "Riverscape" is located on the ba of the Qinhuai River, which is a high-density residential area with a lack of medic services, cultural facilities and public spaces. In order to solve these problems, t project adopts the strategy of the landform building to build a community cent beside the river, with an attempt to not only provide several community services, b also connect the city to the river view. The project "linear utopia" aims at creating linear and continuous multifunctional public space, meant to solve the problem the deficiency of public spaces and private spaces in the dormitory area at Nanji University. By choosing "interior landscape" as the urban strategy, this project tri to organically combine several public spaces with the existing block to create dynamically integrated landscape.

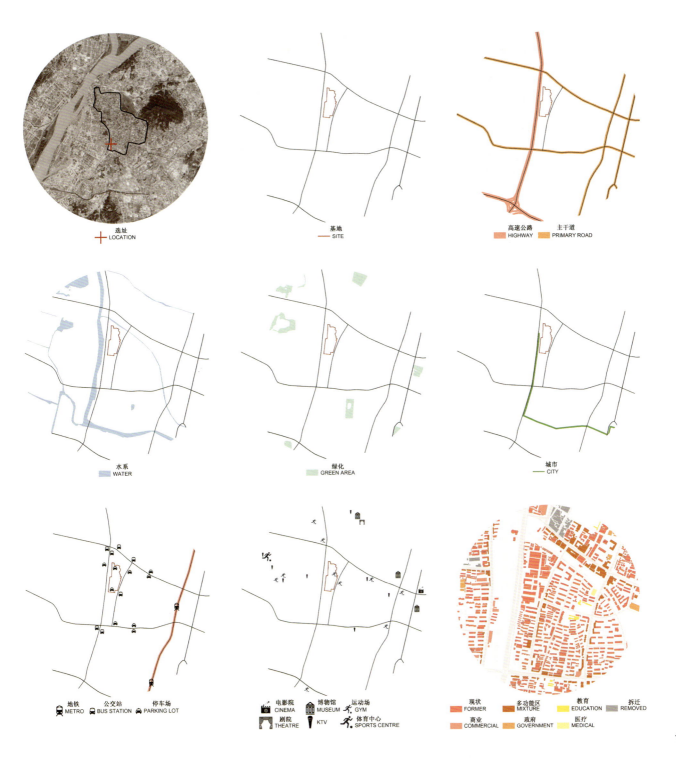

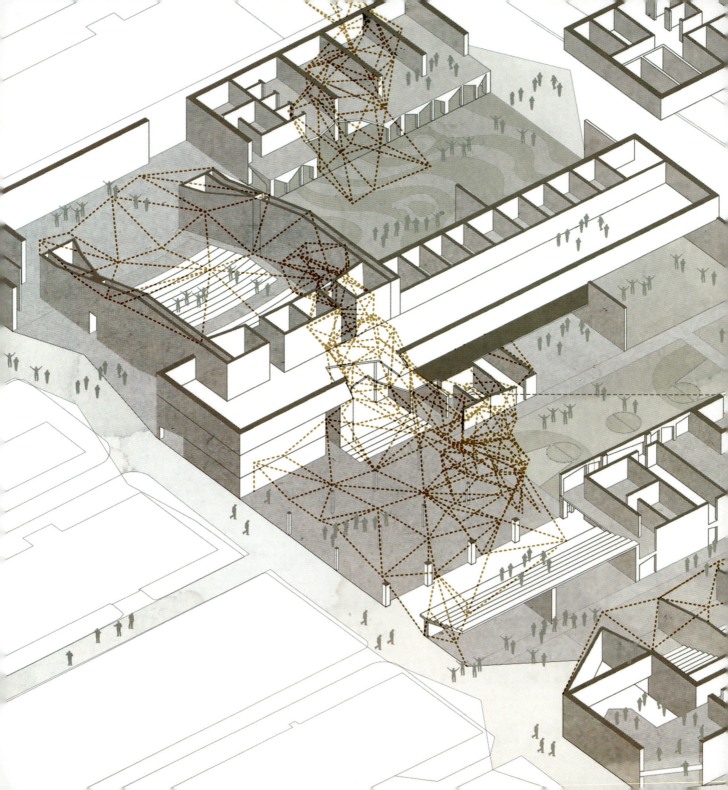

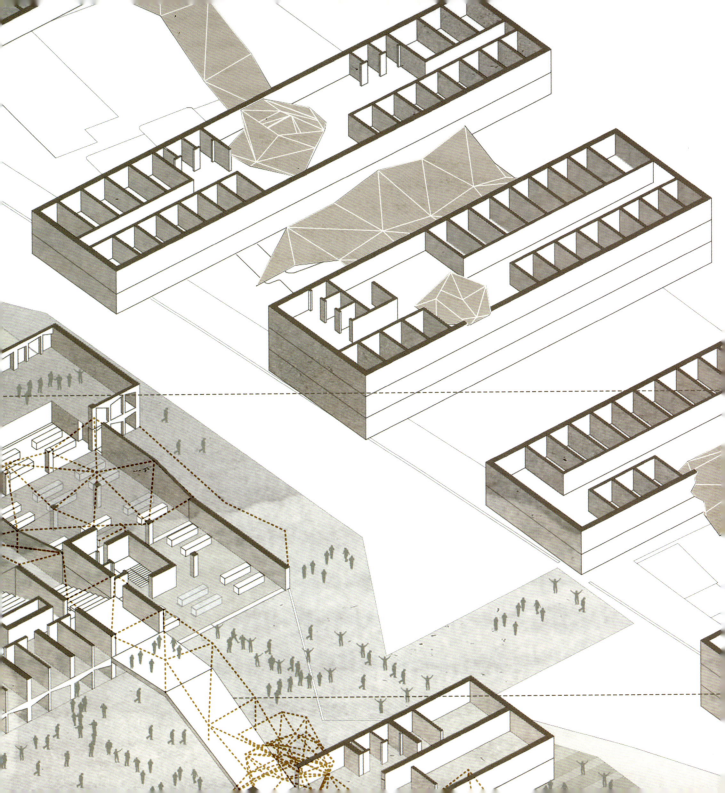

研究生国际教学交流计划课程 IPTP

参数化图解静力学
PARAMETRIC GRAPHIC STATICS WORKSHOP

科朗坦·菲韦 孟宪川

参数化图解静力学设计工作营致力于在设计初期促进建筑与结构的融合。自18世纪建筑学与结构工程专业分工以来，两个学科各自发展迅速，同时也加深了彼此间的隔阂，因此两学科相互融合的议题被不断探索。鉴于建筑师的图形思维方式与结构师的数理思维方式迥异，导致双方难以有效地合作。图形作为媒介的研究方法成为架构两个专业间桥梁的工具。参数化的图解静力学正是其中一种前沿的设计方法。

参数化图解静力学设计工作营为建筑师提供了创新的方法与工具，使建筑师们通过数字化方法，操作形式的内部力流，设计结构的几何形式。工作营首先通过案例介绍基本知识。几何技巧加快了形式的建构，以此来广泛地探索静力平衡状态下的结构形态。这些技巧通过参数化的工具（Rhino 3D和Grasshopper）被应用于设计，有利于在设计早期阶段通过客观的标准对结构进行优化。历史上优秀案例的设计过程也将被逐步地揭示。

每个小组的设计成果由四个部分组成：
(1) 根据设计要求提出初始的概念意向；
(2) 通过不断整合各种设计要素，形成迭代的设计过程；
(3) 利用参数化图解静力学的工具对结构进行优化；
(4) 表达最终的设计结果。

到工作营结束，参与的同学具备一定能力：
(1) 重构设计策略以探索符合低碳要求的结构体系；
(2) 更好地判断给定结构类型的设计/几何的自由度；
(3) 在静态平衡的前提下，针对概念结构体系定制参数化的模型；
(4) 更好地判断如何修改结构的几何形态以增强其结构性能。

This workshop on parametric graphic statics attempts to promote the integration between architecture and structure during the early conceptual phase of the design project. Starting from the 18th century the education of architects and structural engineers divided in two separate directions and high-speed developed separately, the gap between each other was deepened, therefore the issue of integration of two professions has been continuously explored. In view of the different between the graphic thinking model of architects and the mathematical thinking model of engineers, it is difficult for both professions to cooperate effectively. Graphical approaches are potentially very good media to bridge such gap between architects and engineers. Parametric graphic statics is one of such frontier design approaches. This workshop presents innovative methods and tools that give the architects the opportunity to control the design of a structural geometry together with its internal flow of forces. The workshop will start with the basics, introducing the general rules by means of practical examples. Specific methods are then developed with application to various structural materials. Geometric shortcuts are emphasized in order to speed up the graphical construction, and hence the wide exploration of structural arrangements in static equilibrium. Implementations of these methods into a parametric software tool (Rhino 3D and Grasshopper) are addressed alongside with objective criteria that can be used early in the process to optimize the structure. Historical examples of interactive design processes will also be showcased.

The final presentation of each group includes four parts:
(1) Conceptual intension based on the design conditions;
(2) Iterations of the design process following the integration of multiple design factors;
(3) Structural optimization with parametric graphic static tools;
(4) Renderings of the final design.

At the end of the workshop, participants are able to:
(1) Reproduce conceptual design strategies to explore structural systems that have a low-carbon impact;
(2) Better determine the degrees of design/geometric freedom of a given structural typology;
(3) Build tailored parametric explorations of conceptual structural systems in static equilibrium;
(4) Better determine how to modify the geometry of a structure in order to enhance its static behavior.

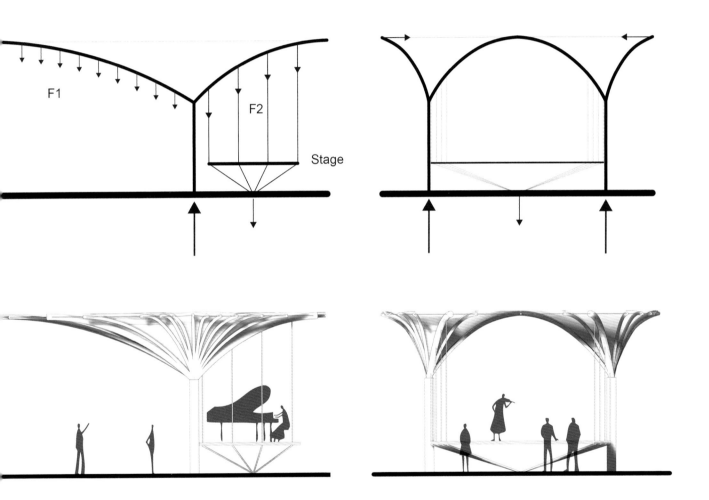

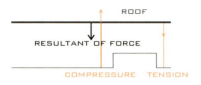
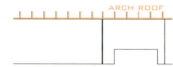
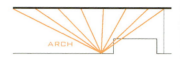

基本结构
1. BASIC STRUCTURE

弧形屋顶
2. ARCH ROOF

拱
3. ARCH

拱顶
4. ARC ROOF

交叉，反作用力小
5. INTERSECTION, SMALLER REACTION

不交叉，易建造
6. NON-INTERSECTION, EASIER TO BUILD

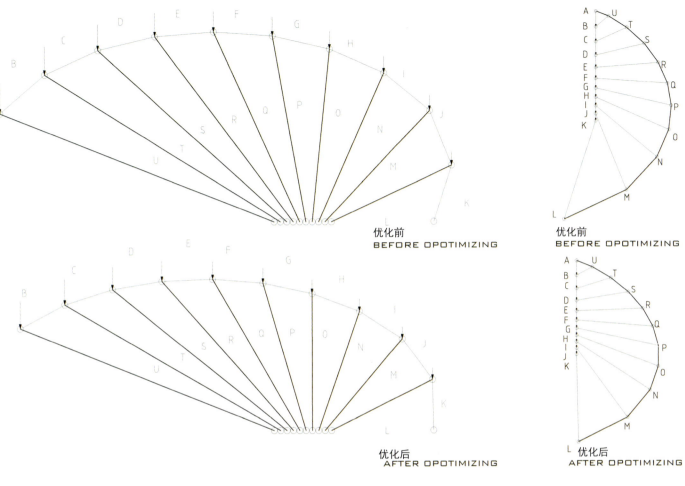

优化前
BEFORE OPOTIMIZING

优化后
AFTER OPOTIMIZING

研究生国际教学交流计划课程 IPTP

环境交互设计
ENVIRONMENTAL INTERACTIVEDESIGN

若泽·阿尔莫多瓦 唐莲 尤伟

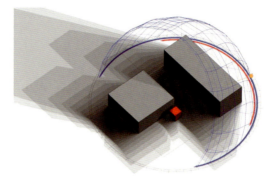

阿尔莫多瓦教授指导的"环境交互设计"工作营向学生介绍了如何设计可持续的建筑以及如何在设计过程中使用不同的环境设计工具。通过对物理环境/场地、形式/体块/朝向、室内构成、外部围合的效果的整体分析和性能模拟,课程提供了跨学科及跨文化的环境来学习和探索新的设计概念以及策略。环境设计的概念和方法将引领整个设计过程。在为期两周的工作和研讨中,学生们在阿尔莫多瓦教授的指导下探索如何分析与评估并改善建筑的环境性能,最终取得较好的设计成果。

课程包括两个部分:讲座环节和工作坊环节,强调从做中学。在讲座环节,阿尔莫多瓦教授引导学生充分理解传统建筑材料、形式、空间组织与环境效能的关系之后,又向同学们介绍了HEED、Climate Consultant、Sefaira、Ecotect等软件,指导学生利用上述分析工具在设计过程中对建筑的天然采光及围护结构传热性能等进行评估与分析。其中HEED和Climate Consultant是加利福尼亚大学洛杉矶分校(UCLA)开发出来的建筑能耗和气候条件分析工具。

在工作坊环节,每个学生被要求选择一个建筑案例作为设计优化对象,先利用Climate Consultant分析建筑所在区域的气候条件,确定该建筑的优化目标,然后利用Sefaira、Ecotect软件来分析建筑案例中存在的天然采光、热工性能问题,最后提出解决方案。阿尔莫多瓦教授在研究的基础上总结了建筑围护结构各立面的设计要点,并教授同学们用图解的方式去思考优化采光及日照问题。

The "Environmental Interactive Design" workshop, directed by professor José ALMODÓVAR introduces students how to design sustainable buildings and to use different environmental design tools. It provides a cross disciplinary and cross cultural environment for learning and exploring new design concepts and strategies through integrated analysis and performance simulations of effects of physical environment and site, form, massing and orientation, internal configuration, external enclosure.

The concept and methods of the environmental design will guide the who design process. In the two-week work and study, under the direction of profess ALMODÓVAR, the students explored how to analyze and improve the buildin environmental performance, and finally achieved better design works.
The course includes two parts: lecture series and workshop that emphasis learning by doing. After guiding students to fully understand the relationsh between traditional architecture's material, configuration, spatial organization a environmental efficiency, professor ALMODÓVAR introduced environmental desig tool, including Home Energy Efficient Design (HEED), Climate Consultant, Sefair Ecotect, and directed the students to carry out building's daylighting and therm performance of the building envelope in the design process, with the help of t analysis tool mentioned above. Among these tools, HEED and Climate Consulta are developed by University of California Los Angeles (UCLA), which can respectiv assess building energy consumption and climatic condition.
In the following teaching process, each student was required to choose a constructio case as the object to optimize. Firstly, using Climate Consultant to analyze t climatic conditions where the building located, and then, the optimal goal of t building should be determined, after that, analyzing the existing problems-natur lighting and the thermal performance of this building-by using Sefaira and Ecotec Finally, the solution was put forward according to the analysis above. On the bas of the research, Professor ALMODÓVAR summarized the main points when desig each facade of the building envelope, and also taught the students to think how optimize the daylighting and sunshine problems in a graphic way.

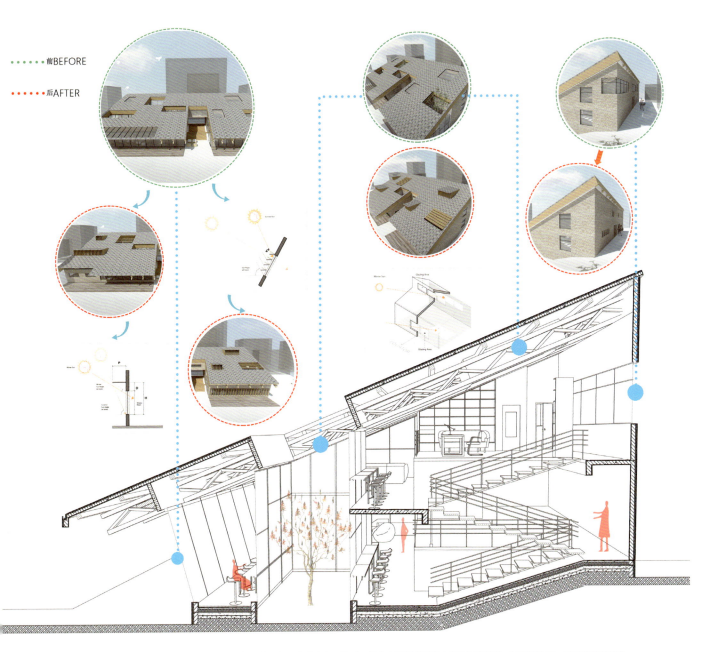

用Climate Consultant分析建筑所在区域的气候条件，确定该建筑的优化目标，然后利用Sefaira、Ecotect软件来分析建筑案例中存在的天然采光、热工性能问题，最后提出解决方案。

Use Climate Consultant to analyze the climatic conditions where the building located, and then, the optimal goal of the building should be determined, after that, analyzing the existing problems-natural lighting and the thermal performance of this building-by using Sefaira and Ecotect. Finally, the solution was put forward according to the analysis above.

Daylighting-Before

天然采光—前

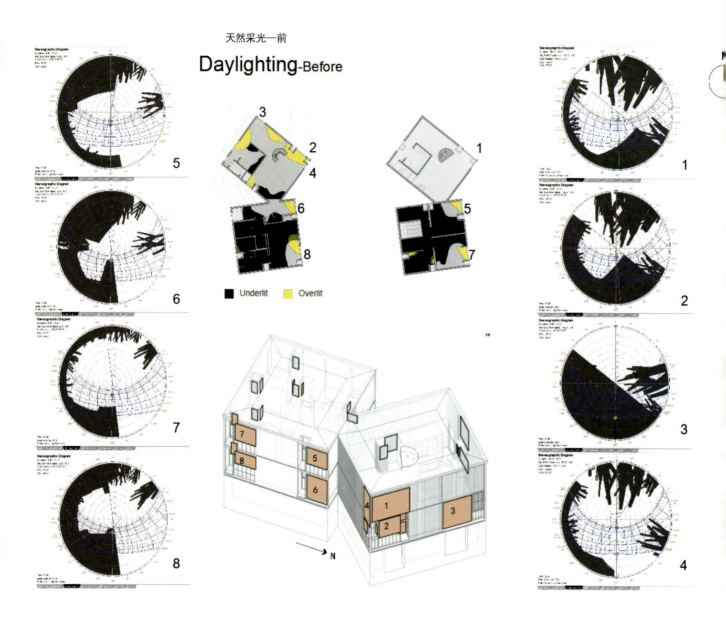

■ Underlit　　■ Overlit

天然采光—后

Daylighting-After

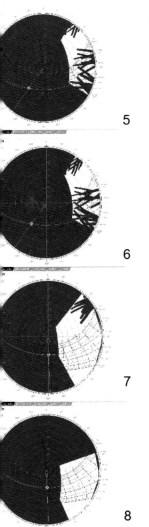

5

6

7

8

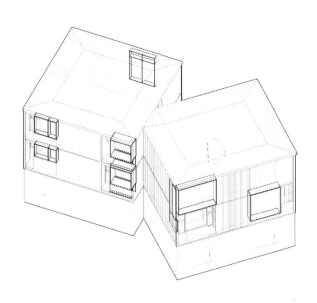

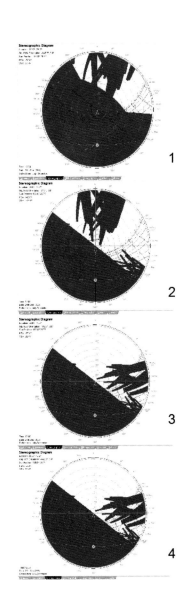

1

2

3

4

研究生国际教学交流计划课程 IPTP

繁荣的后合理化
RE-APPROPRIATE THE BOOM

弗洛里·科萨克 窦平平

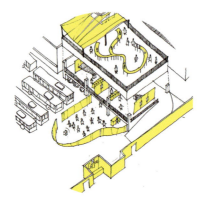

本组工作坊的主要目的是运用一些分析手段和分析工具深刻体会在城市繁荣发展中一些被居民、市民等使用者再利用的空间。

城市设计和规划通常被理解为从上帝视角构想的干预措施：不同尺度的规划视角通常是制定干预策略方法的基础。从上帝的视角观察——通过鸟瞰——可以提供对特定地区的物理和形式方面的强有力的见解，但这种方式忽视了许多贯穿干预措施的问题。例如，从上帝的视角观察使我们完全错过了其他可见事物，诸如占有痕迹、公共空间的实际用途、肌理、材料、修复的状态等。然而，至关重要的是，我们也很难理解该地区的无形和不可见的方面，如其空间的总体氛围、听觉和视觉质量，还有政治压力、文化特征、日常挑战与欲望的迹象。

本工作坊通过进入情境、收集、调查、制图、提案、沟通、反思七大步骤期望能够为同学们提供新的认知城市的视角。

The main purpose of this workshop is to use some analytical methods and tools understand some re-appropriated space which used by residents, citizens and so other users in the boom-developed urban deeply.

Urban design and planning are commonly understood as interventions conceive from above: plan views at different scales are often the basis for devising strate approaches for interventions. Whilst looking at an area from above–through a bird eye view–can offer powerful insights into physical and formal aspects of a particu area, this approach is also generally blind to many issues that should inform c interventions. For instance, looking from above, we completely miss otherwi visible things such as traces of appropriation, actual uses of public space, texture materials, state of repair, and so forth. Crucially, however, we also miss out on gaini an understanding of intangible and non visible aspects of the area such as its gene atmosphere, its audio and visual qualities, but also signs of political tensions, cultu characteristics, daily challenges and desires.

This workshop expects to provide students with new view of cognizing the c through seven steps including situation, gathering, surveying, mapping, proposi communicating, reflecting.

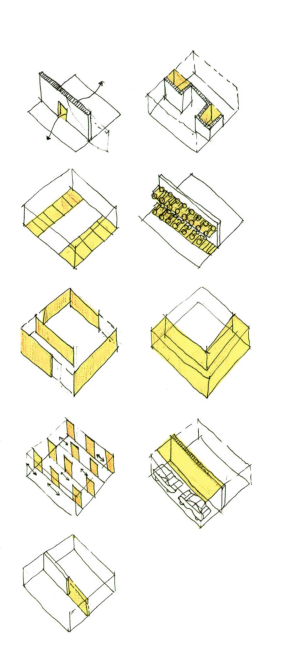

纷杂的场地压缩到实在的墙，再提出适应性的设计原则辐射回整个场地。
Abstract the wall element from a complicated site, survey it and then propose relevant but general strategies to be applied throughout the whole site.

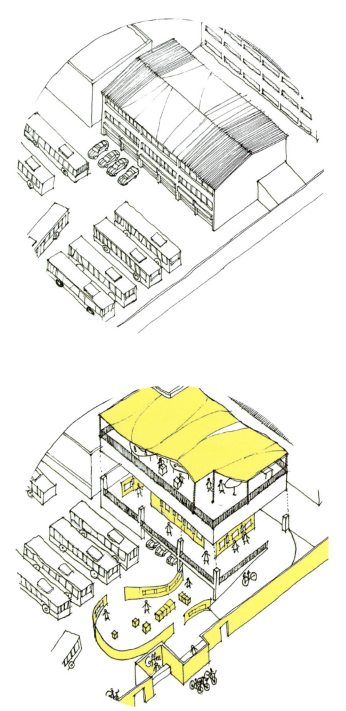

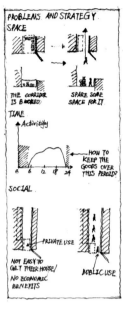
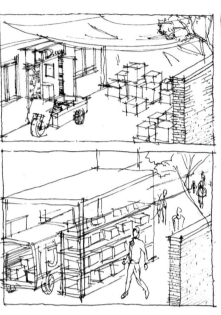
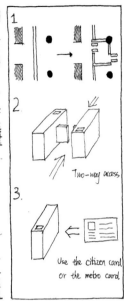
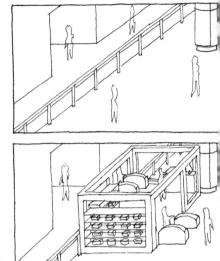
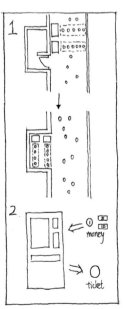
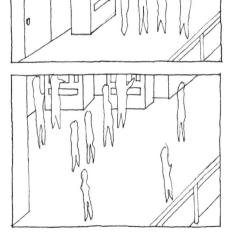
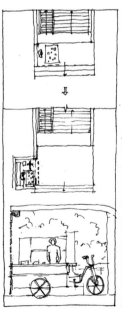

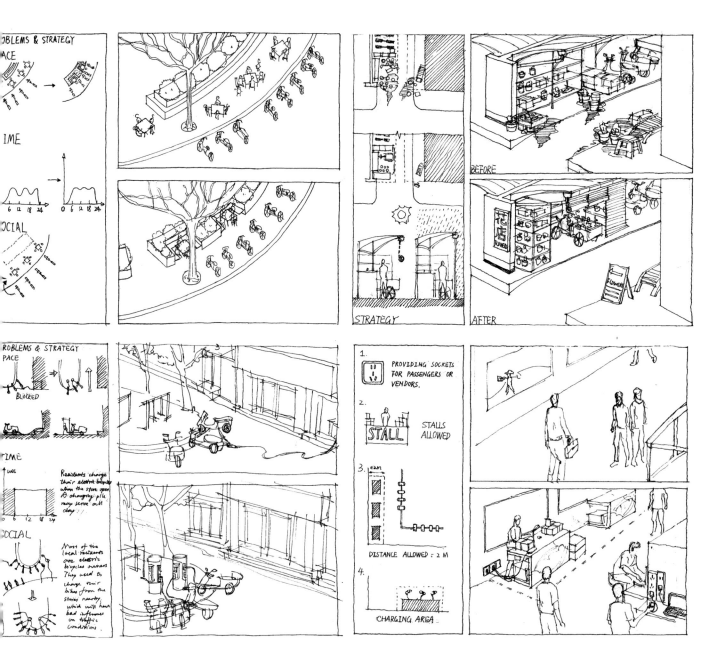

建筑设计课程
ARCHITECTURAL DESIGN COURSES

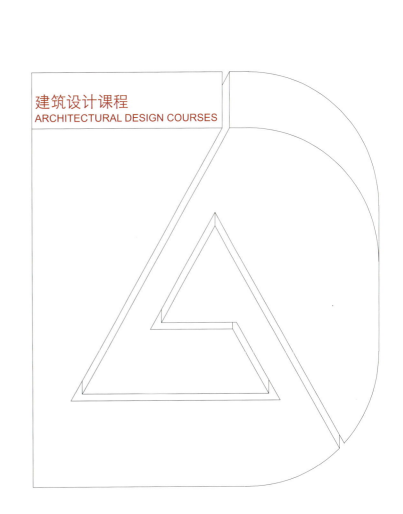

本科一年级
设计基础（一）
· 季鹏

课程类型：必修
学时学分：64学时／2学分

Undergraduate Program 1st Year
BASIC DESIGN 1·JI Peng
Type: Required Course
Study Period and Credits: 64 hours/2 credits

教学目标
　　提升学生感知美、捕捉美和创造美的能力。
研究主题
　　1.看到世界的美。
　　2.看到形式的美。
　　3.理解空间的规则。
　　4.理解使用的规则。
教学内容
　　1.研究对象的素描表达、黑白灰归纳、拼贴表现和陶土烧造。
　　2.针对Zoom in概念的训练与纸构成训练。
　　3.建筑摄影。
　　4.观念与材料的关系。

Training Objective
Improve ability of students to perceive, capture, and create beauty
Research Subject
1. See beauty of the world.
2. See beauty of forms.
3. Understand rules of space.
4. Understand rules of use.
Teaching Content
1. Sketch presentation, black-white-grey induction, collage express and pottery making for research objects.
2. Training on the concept of "Zoom in" and paper construction.
3. Architectural photography.
4. Relations between ideas and materials.

本科一年级
设计基础（二）
· 丁沃沃　鲁安东　唐莲　刘妍

课程类型：必修
学时学分：64学时／2学分

Undergraduate Program 1st Year
BASIC DESIGN 2 · DING Wowo, LU Andong, TANG Lian, LIU Yan
Type: Required Course
Study Period and Credits: 64 hours/2 credits

动作—空间分析
　　通过影像记录人在空间中的动作，挑选关键帧进行动作、尺度、几何与感知分析。目的在于：使学生初步认识身体、尺度与环境的相互影响；学会观察并理解场地；初步认识形式与背后规则的关系；学会发现日常使用中的问题并解决问题；学会使用分析图交流构思。
折纸空间
　　折叠纸板创造空间，用轴测图、拼贴图再现空间。目的在于：使学生初步掌握二维到三维的转化，初步认识图和空间的再现关系；利用单一材料围合复杂空间，初步认识结构与空间的关系；学会用分析图进行表述。
互承的艺术
　　运用互承结构原理，搭建人能进入的覆盖空间。目的在于：初步理解结构知识对于构筑空间的意义；在搭建过程中初步建立材料、节点、造价等概念；强化场地意识（包括朝向、环境、流线等）；学会用分析图进行表述。

Action-space Analysis
Record human's actions in space with video and select frames to conduct analysis on actions, dimensions, geometry a perception. It aims to enable students to preliminarily understand mutual influence among body, dimensions and environment; le how to observe and understand the site; preliminarily understa relations between forms and rules behind them; learn how to find problems in daily use and how to solve these problems; learn how use analysis charts to exchange ideas.
Folding Space
Fold paperboard to create space, and represent space w axonometric drawing and collage. It aims to enable students preliminarily grasp the transformation from 2D to 3D, prelimina understand the reappearance relationship between drawing a space; enclose complex space with single material, and prelimina understand the relations between structure and space; learn how express with analysis charts.
Art of reciprocal structures
Apply principles of reciprocal structures to erect an accessi covered space. It aims to preliminarily understand meaning structure knowledge to construction of space; roughly establish concepts material, joints, and cost during erection; enhance s awareness (including orientation, environment, streamline, etc.) learn expression with analysis charts.

本科二年级

建筑设计基础
刘铨　冷天
课程类型：必修
学时学分：72学时／4学分

Undergraduate Program 2nd Year
ARCHITECTURAL BASIC DESIGN · LIU Quan, LENG Tian
Type: Required Course
Study Period and Credits: 72 hours/4 credits

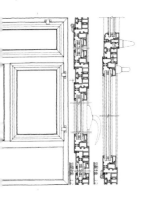

课题内容
　　认知与表达
教学目标
　　本课程是建筑学专业本科生的专业通识基础课程。本课程的任务主要是一方面让新生从专业的角度认知与实体建筑相关的基本知识，如主要建筑构件与材料、基本构造原理、空间尺度、建筑环境等知识；另一方面通过学习运用建筑学的专业表达方法来更好地掌握这些建筑基本知识，为今后深入的专业学习奠定基础。
教学内容
　　1.认知建筑
　　　（1）立面局部测绘
　　　（2）建筑平、剖面测绘
　　　（3）建筑构件测绘
　　2.认知图示
　　　（1）单体建筑图示认知
　　　（2）建筑构件图示认知
　　3.认知环境
　　　（1）街道空间认知
　　　（2）建筑肌理类型认知
　　　（3）地形与植被认知
　　4.专业建筑表达
　　　（1）建筑图纸表达
　　　（2）建筑模型表达
　　　（3）环境分析图表达

Subject Content
Cognition and presentation
Training Objective
The course is the basic course of general professional knowledge for undergraduates of architecture. Task of the course is, on the one hand, allow students to cognize basic knowledge about physical building from an professional perspective, such as main building members and materials, basic constructional principles, spatial dimensions, and building environment; and on the other hand, to better master such basic architectural knowledge through studying application of professional presentation method of architecture, and to lay down solid foundation for future in-depth study of professional knowledge.
Teaching Content
1. Cognizing building
(1) Surveying and drawing of partial elevation
(2) Surveying and drawing plans, profiles of building
(3) Surveying and drawing building members
2. Cognizing drawings
(1) Cognition to drawings of individual building
(2) Cognition to drawings of building component
3. Cognizing environment
(1) Cognition to street space
(2) Cognition to types of building texture
(3) Cognition to terrain and vegetation
4. Professional architectural presentation
(1) Presentation with architectural drawings
(2) Presentation with architectural models
(3) Presentation with environmental analysis charts

本科二年级

建筑设计（一）：老城住宅设计
冷天　刘铨　王丹丹
课程类型：必修
学时学分：54学时／2学分

Undergraduate Program 2nd Year
ARCHITECTURAL DESIGN 1: RESIDENTIAL DESIGN OF OLD TOWN · LENG Tian, LIU Quan, WANG Dandan
Type: Required Course
Study Period and Credits: 54 hours/2 credits

教学目的
　　本次练习的主要任务，是综合运用在建筑设计基础课程中的知识点来推进设计，初步体验用空间形式语言进行设计操作。在设定上，教案希望学生在设计学习开始之初，就关注场地条件、功能与空间、流线组织与人体尺度的紧密关系。
设计要点
　　1. 场地与界面：场地从外部限定了建筑空间的生成条件。本次设计场地是传统老城内真实的建筑地块，面积为100m² 左右，单面或相邻两边临街，周边为1~2层的传统民居。主要要求学生从场地原有界面的角度来考虑设计建筑的形体、布局及其最终的空间视觉感受。
　　2. 功能与空间：本次设计的建筑功能设定为家庭住宅，建筑面积100 m²，建筑高度限制在8m（无地下空间）。
　　3. 流线组织：在给定场地内生成建筑，一方面，内部空间的安排要考虑到与场地界面的关系，如街道界面连续性、出入口的位置等。另一方面，空间的安排要考虑内部流线组织的合理性。
　　4. 人体尺度：在空间形式处理中注意通过图示表达理解空间构成要素与人的空间体验的关系，主要包括尺度感和围合感。

Teaching Objective
This exercise mainly aims to promote the design through comprehensive application of knowledge learned in the basic courses of architectural design, and to preliminarily experience design operation with spatial formal language. The teaching plan hopes that the students can pay attention to the close relationship among site conditions, function and space, streamline organization and the dimension of human figures at the beginning of design learning.
Key Points of Design
1. Site and interface: The site restricts the generating condition of architectural space from the outside. This design site is a real building plot in traditional downtown, covering an area of about 100 m², with frontage on one street or frontages on two streets, surrounded by traditional residences of 1~2 floors. The students are mainly required to consider and design the architectural form and structure, layout and final spatial visual perception from the perspective of original interface of the site.
2. Function and space: The architectural function of this design is defined as family house, with a floorage of 100 m² and a height of 8m (without underground space).
3. Streamline organization: Generate a building in the given site and ensure that, on one hand, the arrangement of internal space takes into account its relationship with the site interface, such as the continuity of street interface, position of entrances and exits, etc.; on the other hand, the spatial arrangement takes into account the reasonability of the internal streamline organization.
4. Dimension of human figures: In the processing of space form, understand the relationship between the space elements and spatial experience of human through graphic expression, mainly including the sense of dimensions and sense of closure.

本科二年级
建筑设计（二）：小型公共建筑设计
· 刘铨 冷天 王丹丹

课程类型：必修
学时学分：54 学时／2 学分

Undergraduate Program 2nd Year
ARCHITECTURAL DESIGN 2: SMALL PUBLIC BUILDING · LIU Quan, LENG Tian, WANG Dandan

Type: Required Course
Study Period and Credits: 54 hours/2 credits

课题内容
　　风景区茶室设计

教学目标
　　本课程承接上一个设计题目，继续训练使用空间形式语言进行设计操作，训练学生进一步掌握"空间形式语言"作为设计。

研究主题
　　1.场地与界面：场地从外部限定建筑空间的生成条件，要求学生从场地水平向界面的限定来考虑设计建筑的形体、布局及其最终的空间视觉感受。
　　要求对坡地界面形态加以形式化的提取和表达，如堆叠、阵列、编制、切片、张拉等，以强化对地形的理解并作为推动后续设计的形式工具。
　　2.功能与空间：建筑面积300 m²，建筑层数2层（无地下空间），其中包括入口门厅、容纳60人的大厅，容纳30人的2~4人雅座若干，以及操作间、储藏间、值班室、卫生间等必要的辅助空间。
　　3.空间的组织：基于地形的形式化表达，寻求适应建筑空间需求的场地改造形式进行高差的处理，把对地形的理解与公共建筑的基本功能空间组织模式及形态塑造联系起来。
　　4.尺度与感知：在空间形式处理中注意通过图示表达理解空间构成要素与人的空间体验的关系，主要包括尺度感和围合感，注意景观朝向问题。

Subject Content
Design of Teahouse in Scenic Spot
Training Objective
This course continues from last design subject, proceeds with th training of design operation with spatial formal language and trai the students to further master the "spatial formal language" f design.
Research Subject
1. The design site is a sloped field within the Zijinshan scenic sp and it requires students to consider the profile, layout and final visu effect of space of the designed building based on limits of horizon interface of the site.
2. The building to be design with functions as a teahouse at scen spot, covers an area of 300 m², consists of two floors, including a entrance hallway, a public hall with seating capacity of 60 peopl several private rooms with seating capacity of 2~4 people whic can serve 30 guests in aggregate, as well as other necessa auxiliary spaces. Meanwhile, design for outdoor space must also b conducted.
3. On basis of formalized expression of topography, try to deal wi height difference in the form of field modification by accommodatin demand of building space, and link understanding of topograph to the space organization patterns and form modelling for bas functions of public building.
4. Pay attention to the understanding of relations between factors space formation and spatial experience of human through graph expression in the process of handling spatial forms.

本科三年级
建筑设计研究（三）：小型公共建筑设计
· 周凌 童滋雨 窦平平

课程类型：必修
学时学分：72 学时／4 学分

Undergraduate Program 3rd Year
ARCHITECTURAL DESIGN 3: SMALL PUBLIC BUILDING · ZHOU Ling, TONG Ziyu, DOU Pingping

Type: Required Course
Study Period and Credits: 72 hours/4 credits

课题内容
　　赛珍珠纪念馆扩建

教学目标
　　此课程训练最基本的建造问题，使学生在学习设计的初始阶段就知道房子如何造起来，深入认识形成建筑的基本条件：结构、材料、构造原理及其应用方法，同时课程也面对场地、环境和功能问题。训练核心是结构、材料、场地。在学习组织功能与场地的同时，强化认识建筑结构、建筑构件、建筑围护等实体要素。
　　文脉：充分考虑校园环境、历史建筑、校园围墙以及现有绿化，需与环境取得良好关系。
　　退让：建筑基底与投影不可超出红线范围。若与主体或相邻建筑连接，需满足防火规范。
　　边界：建筑与环境之间的界面协调，各户之间界面协调。基底分隔物（围墙或绿化等）不超出用地红线。
　　户外空间：扩建部分保持一定的户外空间，户外空间可在地下。
　　地下空间：充分利用地下空间。

教学内容
　　基地内地面最大可建面积约100 m²，地下可建面积200~300 m²，总建筑面积约400~500 m²，建筑地上1层，限高6 m，地下层数层高不限，展示区域200~300 m²，导游没 10 m²，纪念品部 30 m²，茶餐厅 60 m²，厨房区域 >10 m²，另包括 门厅与交通、卫生间。

Subject Content
Extension of Pearl S.Buck's House in Nanjing
Training Objective
This course trains the students to solve the basic construction architecture. Students should learn how to build an architecture a the very beginning of their studying, understand the basic aspec of architectures: the principles and applications of structure, materi and construction. The course also includes the problem of sit environment and function. The keypoints of the course include sit structure and material. Students should strengthen the understandin of physical elements including structures, components and façade while learning to organize the function and site.
Context: The environment, historical building, the edge of th campus and the green belt around the site should be take into consideration. The expansion is expected to have a goo relationship with the surroudings.
Retreat Distance: The new architecture can't beyond the red line Fire protection rule should be complied.
Boundary: Both the boundary between different buildings an between building and environment should be harmonized.
Open Space: Open space should be considered, which permitted t be placed underground.
Underground Space: Underground space should be well used.
Teaching Content
The maximum ground can be used in the base area is about 100 m while underground construction area is about 200~300 m², and the tota floor area of architecture should be about 400~500 m². The architectur should be 1 floor above the ground lower than 6 m. The undergroun levels have no limitation. Exhibition area: 200~300 m², informatio center: 10 m², shop: 30 m², coffee bar: 60 m², kitchen: >10 m lobby & walking space, toilet.

本科三年级
建筑设计（四）：中型公共建筑设计
周凌　童滋雨　窦平平
课程类型：必修
学时/学分：72学时 / 4学分

Undergraduate Program 3rd Year
ARCHITECTURAL DESIGN 4: PUBLIC BUILDING · ZHOU Ling, TONG Ziyu, DOU Pingping
Type: Required Course
Study Period and Credits: 72 hours/4 credits

课题内容
　　傅抱石美术馆
教学目标
　　课程主题是"空间"和"流线"，学习建筑空间组织的技巧和方法，训练空间的效果与表达。空间问题是建筑学的基本问题，课题基于复杂空间组织的训练和学习。从空间秩序入手，安排大空间与小空间，独立空间与重复空间，区分公共与私密空间、服务与被服务空间、开放与封闭空间。同时，空间的串联形成序列，需要有效组织流线，并且充分考虑人在空间中的行为，空间感受。以模型为手段，辅助推敲。设计分阶段做体积、空间、结构、围合等，最终形成一个完整的设计。
教学内容
　　1.空间组织原则：空间组织要有明确特征，有明确意图，概念要清楚。并且满足功能合理、环境协调、流线便捷的要求。注意三种空间：聚散空间（门厅、出入口、走廊）；序列空间（单元空间）；贯通空间（平面和剖面上均需要贯通，内外贯通、左右前后贯通、上下贯通）。
　　2.空间类型：展览陈列空间：3 000 m²；收藏保管空间：700 m²；技术、研究空间：240 m²；行政办公空间：150 m²；休闲服务空间：300 m²；其他空间：传达室：10 m²；设备房：200 m²；交通门厅面积自定；客用、货用电梯各一部；室外停车场。
　　建筑面积不超过5 000 m²，高度不超过18 m。

Subject Content
Fu Baoshi Art Gallery
Training Objective
The course topic is space and circulation, learning techniques and methods of architectural space organization, and training on effect and presentation of space. Space is a basic issue for architecture, and the course is based on training and study on organization of complex spaces. Start from spatial order to arrange large and small spaces, independent space and overlapped space, and to distinguish public and private spaces, serving and served spaces, open and closed spaces. Meanwhile, linking spaces to shape sequence requires effective organizational circulation, as well as full consideration of behaviors, spatial feeling of people in space. Use models as means to assist deliberation. Design stages include volume, space, structure, and enclosure, and shape a complete design in the end.
Teaching Content
1. Space organization principles: Space organization requires distinctive characteristics, explicit intention, and clear concepts. It shall also meet the requirements of reasonable functions, coordinated environment, and convenient circulation. Attention shall be paid to three types of space: converging and diverging space (hallway, entrance and exit, corridor); sequence space (unit space); connecting space (connecting spaces are required on plans and profiles, internal-and-external connection, left-and-right, front-and-rear connections, up-and-down connection).
2. Space type: exhibition & showcase space: 3 000 m²; collection & storage space: 700 m²; technical, research space: 240 m²; administrative office space: 150 m²; leisure service space: 300 m²; other spaces: janitor's room: 10 m²; equipment room: 200 m²; area of traffic hallway to be determined; one guest elevator and one goods elevator; outdoor parking lot.
The floor area shall not exceed 5 000 m², and the height shall not exceed 18 m.

本科三年级
建筑设计（五 +六）：大型公共建筑设计
华晓宁　钟华颖　王铠
课程类型：必修
学时/学分：144 学时 / 8 学分

Undergraduate Program 3rd Year
ARCHITECTURAL DESIGN 5&6: COMPLEX BUILDING · HUA Xiaoning, ZHONG Huaying, WANG Kai
Type: Required Course
Study Period and Credits: 144 hours/8 credits

课题内容
　　城市建筑：社区中心
研究主题
　　实与空：关注城市中建筑实体与空间的相互定义、相互显现，将以往习惯上对于建筑本体的过度关注拓展到对于"之间"的空间的关注。
　　内与外：进一步突破"自身"与"他者"之间的界限，将个体建筑的空间与城市空间视为一个连续统，建筑空间即城市空间的延续，城市空间亦即建筑空间的拓展，两者时刻在对话、互动和融合。
　　层与流：不同类型的人和物的行为与流动是所有城市与建筑空间的基本框架，当代大都市中不同的流线在不同的高度上层叠交织，构成一个复杂的多维城市。必须首先关注行为和流线的组织，由此才生发出空间的系统和形态。
　　轴与界：城市纷繁复杂的形态表象之后隐含着秩序和控制，并将成为新的形态介入。
教学内容
　　在用地上布置社区商业中心（约 15 000 m²）、社区文体活动中心（约 8 000 m²），并生成相应的城市外部公共空间。

Subject Content
Urban Buildings-Design of Community Business Center and Activity Center
Research Subject
Entity and space: Pay attention to mutual definition, mutual representation of architectural entity and space in cities, and extend traditional excessive attention to the building itself to the space "among them".
Interior and exterior: Further break through the boundary between "self" and "others", and consider space of individual buildings and urban space as a continuum.
Stack and flow: Behaviors and flows of different types of people and objects are the basic framework of all urban and building spaces, and a complex multi-dimensional city is formed by stacking up and interweaving of different flow lines at different altitudes in modern metropolis. We must pay attention to organization of behaviors and flow lines first, and then can generate system and morphology of space.
Axis and boundaries: Order and control are concealed behind the morphologic appearance of complexity of cities, which will be involved as new forms.
Teaching Content
Lay out community commercial center (about 15 000 m²) and community recreational and sports activities center (about 8 000 m²) on the land, and generate associated outdoor urban public spaces.

本科四年级
建筑设计（七）高层建筑设计
吉国华　胡友培　尹航
课程类型：必修
学时学分：72学时／4学分

Undergraduate Program 4th Year
ARCHITECTURAL DESIGN 7: HIGH-RISING BUILDING · JI Guohua, HU Youpei, YIN Hang
Type: Required Course
Study Period and Credits: 72 hours/4 credits

课题内容
　　高层办公楼设计
教学目标
　　高层办公建筑设计涉及城市、空间、形体、结构、设备、材料、消防等方面内容，是一项较复杂与综合的任务。本课题采取贴近真实实践的视角，教学重点与目标是帮助学生理解、消化所涉及各方面知识，提高综合运用并创造性解决问题的技能。
教学内容
　　建筑容积率≤5.6，建筑限高≤100 m，裙房高度≤24 m，建筑密度≤40%。需规划合理流线，避免形成交通拥堵。
　　高层部分为办公楼，设计应兼顾各种办公空间形式。裙房设置会议中心，须设置400人报告厅1个，200人报告厅2个，100人报告厅4个，其他各种会议形式的中小型会议室若干，以及咖啡茶室、休息厅、服务用房等。会议中心应可独立对外使用。机动车交通独立设置，不得进入校内道路系统。地下部分主要为车库和设备用房。

Subject Content
Design of High-rise Office Building
Training Objective
Design of the high-rise office building is a complicated comprehensive task, involving city, space, form, struct equipment, materials and fire control. From a perspective c to the practice, this course focuses on and aims at hel students understand and grasp the knowledge of the ab mentioned aspects and improving their skills of integrated and creatively solving problems.
Teaching Content
Building plot ratio ≤5.6, building height limit≤100 m, an height ≤ 24 m, building density ≤40%. Reasonable circula must be planned to avoid traffic jam.
The high-rise part is an office building, so the design n give consideration to various forms of office space. The ar is a conference center, which must include 1 lecture ha 400 seating capacity, 2 lecture halls of 200 seating capa 4 lecture halls of 100 seating capacity, several small medium-sized meeting rooms for various meetings, and co & tea room, lobby, and service quarter. The conference ce shall be separated and available for external usage. M vehicle traffic routes must be separated from road sys within the campus. The underground part is mainly for gar and equipment room.

本科四年级
建筑设计（八）城市设计
吉国华　胡友培　尹航
课程类型：必修
学时学分：72学时／4学分

Undergraduate Program 4th Year
ARCHITECTURAL DESIGN 8: URBAN DESIGN · JI Guohua, HU Youpei, YIN Hang
Type: Required Course
Study Period and Credits: 72 hours/4 credits

课题内容
　　南京碑亭巷地块旧城更新城市设计
教学目标
　　1.着重训练城市空间场所的创造能力，通过体验认知城市公共开放空间与城市日常生活场所的关联，运用景观环境的策略创造城市空间的特征。
　　2.熟练掌握城市设计的方法，熟悉从宏观整体层面处理不同尺度空间的能力，并有效地进行图纸表达。
　　3.理解城市更新的概念和价值；通过分析理解城市交通、城市设施在城市体系中的作用。
　　4.多人小组合伙，培养团队合作意识和分工协作的工作方式。
教学内容
　　1.设计地块位于南京市玄武区，总用地约为6.20hm²。地块内国民大会堂旧址、国立美术陈列馆旧址和北侧邮政局大楼可保留，其余地块均需进行更新。地块周边有丰富的博物馆、民国建筑等文化资源，设计应对周边文化环境起到进一步提升作用。地块周边用地情况复杂，设计中需考虑与周边现状的相互影响。
　　2.本次设计的总容积率指标为2.0~2.5，建筑退让、日照等均按相关法规执行。
　　3.碑亭巷、石婆婆庵需保留，碑亭巷和太平北路之间现状道路可根据设计进行位置或线形调整。
　　4.地下空间除满足单一地块建筑配建的停车需求外，应综合考虑地上、地下城市一体化设计与综合开发。

Subject Content
Urban Design for Old Town Renovation of the Land Parcel of Be Lane in Nanjing
Training Objective
1. Emphasize the training on ability of creating urban spatial pla and create features of urban space with the strategy of landsc environment through experiencing and perceiving the links betw urban public spaces and urban daily living places.
2. Master methodology of urban design, grasp the ability of hanc spaces of different dimension at macro and integral level, achieve effective representation with drawings.
3. Understand the concept and value of urban renovat understand the role of urban traffic, urban facilities in the ur system through analysis.
4. Form partnership with several group members to cultiv awareness of teamwork and the working mode of collaboration.
Teaching Content
1.Land parcel of the design is located in Xuanwu District, Nanj covering an area of 6.20hm² approximately. In the site, the for National Assembly Hall, the former National Art Gallery and post office building at north side may be retained, other part the land parcel need to be renovated. There are abundant cult resources such as museums and buildings constructed in the pe of the Republic of China around the land parcel, so the des should further improve the cultural environment around it. Land conditions around the land parcel is very complicated, so mu influence with surrounding existing conditions must be taken account in the design.
2.Gross plot ratio of the design is 2.0~2.5, and building setback sunlight value shall comply with relevant laws and regulations.
3.Beiting Lane and Shipopo Nunnery are to be retained, and loca and route of the existing road between Beiting Land and Ne Taiping Road may be adjusted according to the design.
4.For underground space, besides meeting the associated par demand of buildings on the single land parcel, overall considera shall be taken for integrated design and comprehensive developm of urban spaces above and under the ground.

本科四年级
毕业设计
赵辰
课程类型：必修
学时学分：1 学期 /0.75 学分

Undergraduate Program 4th Year
THESIS PROJECT · ZHAO Chen
Type: Required Course
Study Period and Credits: 1 term /0.75 credit

课题内容
锦屏复兴计划，闽东北山区低碳生态型度假村落规划项目

教学内容
锦屏位于福建东北部山区的南平市政和县生态、人文资源都十分优秀的地域，因交通缘故而长期受到发展的限制。随着高速公路与铁路的发展将迅速得到新的发展机遇，而新的发展必须适应新的低碳、生态型的发展模式，并同时充分保护与发挥地域历史文化与生态的优势。这正是新时代对建筑与乡村规划设计的一种挑战。

教学目标
掌握建筑设计基本的技能与知识（测绘、建模、调研、分析），并能对特定的地域和历史建筑进行深入的设计研究（内容策划、建筑结构、构造），根据社会发展的需求，提出改造和创造的可能。在选定的村落之现状研究的基础上，进行村落景观空间的整体规划。并且，选择相关重点区域与建筑，进行专项的建筑设计。

Subject Content
Jinping Revival, Planning Project of Eco-Low Carbon Holiday Village in Mountainous Area of Northeast Fujian

Teaching Content
Jinping is located in Zhenghe County, Nanping City in the northeastern mountainous area of Fujian, as a region with excellent ecological and cultural resource, while the development is limited for a long time due to traffic. New development opportunity will be rapidly obtained with the development of expressway and railway, new development must adapt to new low-carbon, ecological development mode, and fully protect and give play to the advantage of regional history, culture and ecology. This is a challenge to architectural and rural planning design in the new time.

Training Objective
Master the basic skill and knowledge of architectural design (surveying and mapping, modeling, survey, analysis), can have in-depth design research (content planning, building structure, construction) of specific region and historical architecture, according to the requirement of social development, propose possibility of reform and creation. On the basis of researching current situation of selected village, conduct overall planning of village landscape space. In addition, select relevant key region and architecture for specific architectural design.

本科四年级
毕业设计
郜志
课程类型：必修
学时学分：1 学期 /0.75 学分

Undergraduate Program 4th Year
THESIS PROJECT · GAO Zhi
Type: Required Course
Study Period and Credits: 1 term /0.75 credit

课题内容
绿色建筑技术研究：太阳能通风技术及设计策略

教学内容
基于可持续发展的目标，绿色建筑技术是未来建筑技术科学研究中的重要方向。太阳能是新能源与可再生能源中最引人注目的清洁能源，目前普遍认为充分利用太阳能是各种节能途径中潜力最大、最为直接有效的方式，是缓解能源紧张、解决社会经济发展与能源供应不足矛盾的最有效措施之一。太阳能一体化建筑技术将低碳、环保、生态等概念融入太阳能有效利用及绿色建筑技术中，是中低温热能高效利用研究领域的一项重要研究内容。本毕业论文题目旨在研究典型气候与环境参数下，采用太阳能通风技术的建筑环境系统及其气流组织与热量、湿度的整体传播规律及相互关联性，并通过优化太阳能通风系统的结构尺寸、位置及材料等方法与手段，实现最大限度地利用太阳能资源并兼顾人体舒适性与室内空气品质，最终为太阳能一体化建筑设计提供理论依据及实践策略。

教学目标
本课题将开展以太阳能集热与通风技术为主题的专题研究，成果以毕业论文的形式呈现。

Subject Content
Research on Green Building Technology: Solar-powered Ventilation Technology and Design Strategy

Teaching Content
Basing on the objective of sustainable development, green building technology is the important direction in future building technology scientific research. Solar energy is the most remarkable clean energy in new energy and renewable energy, and currently it is generally believed that to fully utilize solar energy is one of the greatest and most direct and effective energy saving approaches, and one of the most effective measure to relieve energy shortage, solve contradiction of social and economic development and insufficient energy supply. Solar integrated building technology merges low carbon, environmental protection, ecology, etc. into the effective utilization of solar energy and green building technology, as an important research content in the field of researching on efficient utilization of medium and low temperature thermal energy. The subject of the thesis aims to research typical climate and environmental parameter, building environment system that adopts solar ventilation technology, overall transmission rule and correlation of air flow organization and heat, humidity, through optimizing structure dimension, position and material, etc. of solar ventilation system, utilize solar energy to the greatest extent and consider human body comfort and indoor air quality, finally provide theoretical basis and practical policy to solar integrated architectural design.

Training Objective
In this subject, special research themed in solar collection and ventilation technology will be conducted, and the result will be presented in the form of thesis.

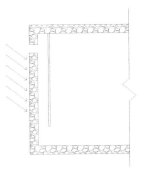

本科四年级
毕业设计
· 华晓宁

课程类型：必修
学时学分：1 学期 /0.75 学分

Undergraduate Program 4th Year
THESIS PROJECT · HUA Xiaoning
Type: Required Course
Study Period and Credits: 1 term /0.75 credit

课题内容
性能导向的乡村复兴：南京市高淳武家嘴村更新设计
教学内容
乡村更新与复兴是当前我国建筑学所面临的主要课题之一。基于性能导向的设计是乡村复兴的重要策略，其目标是引领乡村走向一个可持续发展的未来。性能导向策略是以建筑性能为出发点、以定量的科学模拟与测量为数据支撑、以规范化的相关建筑标准与评估体系为准则的精确地进行建筑设计的方法。
教学目标
本课题将开展基于性能导向的乡村复兴与建筑设计研究，成果以项目设计的形式呈现。

Subject Content
Performance Oriented Rural Renovation: Renovation Wujiazui Village in Gaochun district of Nanjing
Teaching Content
Rural renewal and revival is one of the main subjects th Chinese architecture is facing. Performance oriented desi is the important policy of rural revival, with the purpose leading rural area to a sustainable future. Performance orient policy is the method starting with building performance, w quantitative scientific simulation and measurement as da support, normative building standard and evaluation system rule for precise architectural design.
Training Objective
In this subject, research on performance oriented rural revi and architectural design will be conducted, and the result v be presented in the form of project design.

本科四年级
毕业设计
· 周凌

课程类型：必修
学时学分：1 学期 /0.75 学分

Undergraduate Program 4th Year
THESIS PROJECT · ZHOU Ling
Type: Required Course
Study Period and Credits: 1 term /0.75 credit

课题内容
乡村再造
教学目标
中国传统村镇正在消失，其速度与中国城市化速度成正比，传统农耕文化、传统手工艺、传统价值观处于消失弱化的边缘。本课题着重研究中国传统村镇保护更新的议题。学生通过调研和规划设计，了解传统村镇与传统建筑文化，学习规划知识，训练建筑设计技巧。
教学内容
毕业设计将以江苏南京、浙江的松阳地区乡村营建为主题进行调研、测绘、改造设计与实践。
题目1：高淳武家嘴村乡村更新项目
南京高淳地区水系发达，阡陌纵横，村落依水而建。武家嘴村毗邻石臼湖，保持着完整的传统村落结构，村中有两座祠堂，一个中心广场。房屋环绕村中水塘而建，还有数量不少的传统民宅。由于产业区位变化，村落空心化严重，目前村中只有部分老人居住。政府和企业拟将村落改造为一个文旅小镇，包含文化展示、艺术公园、乡村教育培训、餐饮配套、艺术酒店等在内的特色文化小镇，将吸引来自四面八方的旅游者、商家、文化学者、农品经销商，不同的人群都能在这里找到自己需要的情感、物产和交流方式。研究将以乡村更新改造为题，探讨乡村遗产再利用的技术和社会问题。
题目2：浙江松阳地区传统村落改造更新设计
浙江松阳地区保留了大量传统村落，多数失去了原有功能和活力，有待复兴。榔树村海拔约600m，离松阳城区大约一个钟的车程。20多棵古树背靠大山，环绕整个村子。村口有古庙，相传已有千年，村口的水系是松阳主要水源的发源地，溪水上建有石板路和石桥。这里不但有参天古木，还有涓涓细流，黄土房错落有致依山而建，山上有成片的竹林，但因时代变迁，房屋年久失修。更新改造目标是把遗失的美好记忆保留下来，将破败的房屋修缮利用起来。

Subject Content
Rural Reconstruction
Teaching Content
Traditional Chinese villages are disappearing in dire proportion to Chinese urbanization, traditional farming cultu traditional handicraft, traditional value are on the ver of disappearing and weakening. This subject focuses researching the protection and renewal of Chinese traditi village. Students understand traditional village and traditio building culture, learn planning knowledge, train architectu design skill through survey and planning design.
Training Objective
The graduation design will have survey, surveying a mapping, reform design and practice with the rural building Nanjing, Jiangsu, or Songyang, Zhejiang as the theme.
Subject 1: Gaochun Wujiazui Village renewal project
Nanjing Gaochun has developed water system, with numero footpaths in the field, and the village is built aside the water. Wujia Village is adjacent to Shijiu Lake, maintains complete traditio village structure, and there are two ancestral halls and one cen square in the village. Houses are built around the water pond in t village, and there are many traditional residences. Due to the chan of industrial location, the hollowing situation of the village is seriou currently only few older people live in the village. The governme and enterprises plan to reform the village into a cultural tourism tow including culture display, art park, rural education training, caterin art hotel, etc., which will attract tourists, merchants, cultural schola agricultural product dealers, and different groups can find emotio material and exchange they need here. The research will be theme in rural renewal and reform, and discuss the technology of reutilizi rural heritage and social issue.
Subject 2: Zhejiang Songyang traditional village reform and renew design
A lot of traditional villages are reserved in Zhejiang Songyan most have lost original function and vigor and are to be revive Langshu Village, with the altitude of about 600m, is about 15m drive to Songyang urban area. More than twenty ancient trees ba on the mountain, circle the whole village. There is ancient temp at the entrance of the village, with the history of a thousand yea according to the legend, the water system at the village entrance the origin of main water source of Songyang, and stone path an bridge are built on the creek. There is towering ancient tree an trickling stream, loess houses are built against the mountain in ord there are patches of bamboo forests on the mountain, howeve the houses have been for long years out of repair. The purpose renewal and reform is to preserve the lost beautiful memory an repair and utilize the damaged houses.

本科四年级
毕业设计 · 吉国华
课程类型：必修
学时学分：1 学期 /0.75 学分

Undergraduate Program 4th Year
THESIS PROJECT · JI Guohua
Type: Required Course
Study Period and Credits: 1 term /0.75 credit

课题内容
　　数字化设计与建造
教学内容
　　以"自行车棚"和"休息亭"为题，研究参数化设计在具体项目中的应用，并利用 CNC 机床、3D 打印、机械臂进行模型制作和加工建造。
教学目标
　　学习和掌握参数化设计的工具（Grasshopper 及脚本编程）、参数化设计的方法和数字加工基本技术。在此基础上，完成相关的设计任务，并进行实际搭建。

Subject Content
Digital Design & Construction
Teaching Content
Theme in "bicycle shed" and "half-way house", research the application of parameter design in specific project, utilize CNC machine tool, 3D printing, mechanical arm for modeling and processing.
Training Objective
Learn and master parameter design tool (Grasshopper and script programming), method of parameter design and digital processing basic technology. On this basis, complete relevant design task and actually build.

本科四年级
毕业设计 · 丁沃沃
课程类型：必修
学时学分：1 学期 /0.75 学分

Undergraduate Program 4th Year
THESIS PROJECT · DING Wowo
Type: Required Course
Study Period and Credits: 1 term /0.75 credit

课题内容
　　福建长汀历史文化名城：城市更新与建筑设计
教学目标
　　基于历史文化名城保护与更新的城市设计的真实项目，本毕业设计涵盖了历史文化知识、城市知识、典型民居类型、建筑设计与建造等训练计划，旨在通过训练学习科学有效的调研与分析方法、多重限定下的建筑设计、面向建造的真实问题和城市设计的现实意义。在将本科所学知识融会贯通的基础上，理解设计与研究的关系和研究对于设计的价值。
教学内容
　　长汀县位于福建省西南部闽赣边境，依卧龙山而傍汀江，城内保存了众多的寺庙、祠堂和传统大宅院，既有众多的传统木构建筑和夯土建筑，又有民国时期的闽南洋房，1994年被评为国家历史文化名城。历史上，长汀县被称为客家首府，是客家文化重要的聚集地。同时，长汀又是重要的红色根据地，共和国建国时期的主要领导人都曾在此地逗留或小住，古城内保留着多处红色文化遗址。古城周围群山环绕，所以又被评为国家级生态名城。虽然古城拥有独特的旅游资源，但是近年来经济发展的诉求使古城面临着巨大的压力，保护古城和古建筑早已不是技术问题，单纯的保护早已使城市不堪重负，以保护为目的的城市设计被认为是在保护的基础上给城市带来活力的有效途径，而建筑设计是完成目标的最终手段。
　　本课题以长汀古城历史街区为建筑设计研究范围，通过调研和访谈理解设计问题，通过测绘和分析学习传统建筑的类型和优势以及地方建造的工法。阅读文献资料相关理论，并通过具体的设计研究与实验将所学转化为学理层面的知识和设计方法。

Subject Content
Fujian Changting Famous Historic and Cultural City: Urban Renewal and Architectural Design
Teaching Objective
Real urban design project based on protection and renewal of famous historic and cultural city, the graduation design covers knowledge of history and culture, urban knowledge, typical residence type, architectural design and building, etc., aims to learn scientific and effective survey and analysis method trough training, architectural design under multiple limits, real building problem and real meaning of urban design. On the basis of digesting the knowledge learned, understand the relationship of design and research and the value of research to design.
Teaching Content
Changting County is located at the border of Fujian and Jiangxi in the southwest of Fujian, against Wolong Mountain and aside Tingjiang River, many temples, ancestral halls and traditional mansions are preserved in the city, including traditional wooden building and rammed earth building, and South Fujian western style houses during the Republic of China, and the city was assessed as national famous historic and cultural city in 1994. In history, Changting County was known as the capital of the Hakkas, as the important gathering place of Hakka culture. At the same time, Changting was important red base, where the main leaders during the founding of the Republic stayed or temporarily lived, and many red cultural relics are preserved in the ancient city. The ancient city is circled by mountains, and is known as national famous ecological city. Although the ancient city owns unique tourism resource, the appeal of economic development in recent years has made the ancient city face huge pressure, protection of ancient city and ancient architecture has not been a technical issue, the pure protection has made the city over-burdened, protection-purpose urban design is believed to be the effective approach to bring the city with vigor on the basis of protection, and architectural design is the final means of realizing the objective.
In this subject, the historic block of Changting ancient city is the research scope of architectural design, and design issue is understood through survey and interview. Type and advantage of traditional architecture and local construction method are learned through mapping, surveying and analysis. Read relevant theory of literatures, and transform what is learned into knowledge and design method on the level of theory through specific design research and experiment.

研究生一年级
建筑设计研究（一）：基本设计
· 傅筱
课程类型：必修
学时学分：40 学时／2 学分

Graduate Program 1st Year
DESIGN STUDIO 1: BASIC DESIGN · FU Xiao
Type: Required Course
Study Period and Credits: 40 hours/2 credits

课题内容
　　宅基地住宅设计
教学目标
　　课程从 "场地、空间、功能、经济性" 等建筑的基本问题出发，通过宅基地住宅设计，训练学生对建筑逻辑性的认知，并让学生理解有品质的设计是以基本问题为基础的。
研究主题
　　设计的逻辑思维
教学内容
　　在 A、B 两块宅基地内任选一块进行住宅设计。

Subject Content
Homestead Housing Design
Training Objective
The course starts from fundamental issues of architecture su as "site, space, function, and economical efficiency", aims train students to cognize architectural logics, and allow them understand that quality design is based on such fundamen issues.
Research Subject
Logical Thinking of Design
Teaching Content
Select one from two homesteads A and B and conduct housi design.

研究生一年级
建筑设计研究（一）：基本设计
· 张雷
课程类型：必修
学时学分：40 学时／2 学分

Graduate Program 1st Year
DESIGN STUDIO 1: BASIC DESIGN · ZHANG Lei
Type: Required Course
Study Period and Credits: 40 hours/2 credits

课题内容
　　传统乡村聚落复兴研究
教学目标
　　课程从 "环境" "空间" "场所" 与 "建造" 等基本的建筑问题出发，对乡村聚落肌理、建筑类型及其生活方式进行分析研究，通过功能置换后的空间再利用，从建筑与基地、空间与活动、材料与实施等关系入手，强化设计问题的分析，强调准确的专业性表达。通过设计训练，达到对地域文化以及建筑设计过程与方法的基本认识与理解。
研究主题
　　乡土聚落／民居类型／空间再利用／建筑更新／建造逻辑
教学内容
　　对选定的乡村聚落进行调研，研究功能置换和整修改造的方法和策略，促进乡村传统村落的复兴。

Subject Content
Research on Revitalization of Traditional Rural Settlements
Training Objective
This course starts with basic architectural problems lik "environment" "space" "site" and "construction", analyze and studies the texture, architectural type and life style rural settlement, strengthens the analysis of design problem and emphasizes accurate professional expression from th relationship between building and base, space and activitie and material and implementation through spatial reus after function replacement, to obtain basic knowledge ar understanding of the regional culture as well as the proces and methods of architectural design.
Research Subject
Rural settlement / types of folk house / reutilization of space building renovation / constructional logic
Teaching Content
Conduct investigation and research on selected rur settlement, study methodology and strategy of functio replacement and renovation and improvement, and promo revitalization of traditional rural villages.

生一年级
筑设计研究（一）：概念设计
唐克扬
程类型：必修
学分：40 学时／2 学分

duate Program 1st Year
SIGN STUDIO 1: CONCEPTUAL
SIGN · TANG Keyang
e: Required Course
dy Period and Credits: 40 hours/2 credits

课题内容
　　日常空间的创新
研究主题
　　日常和创新这对概念首先来自于它们的社会学含义，但它们的并举也和建筑学中的一些基本概念相关，比如结构设计里的"常规"和"非正式"，空间的预设"程序"和灵活"使用"。课程将以研讨结合作业的形式，使得学生们完成一个由理念到实现、由基本工作模型到外化与表现的纵贯面。
教学内容
　　学生们需要在教师指导下选择一个南京大学校内的日常空间，根据个人兴趣和内容分配的综合需要，这个空间可能是室内、建筑单体或城市景观三种类型之一。学生们需要熟悉基本的相关理论、经典案例和南京大学校园建筑的历史事实，并在第一周做一个类似于"描红"的案例分析。概念设计将首先讨论面向对象的特定设计工具与设计方法论，然后学生需要根据相近或者互补的日常空间功能组成群组，使用类似或者协同的表现手段进行设计图纸和模型的制作，最终在一个设定的场景中检验设计的成果。每个设计工作组由四位同学组成，各自完成一套完整的练习，并协同合作进行课程汇报。

Subject Content
Innovation of Daily Space
Research Subject
The concept of daily and Innovation is from their sociologic connotation, however, their concurrence is related to some basic concept in architecture, such as "regular" and "informal" in structure design, space preset "procedure" and flexible "use". The course will make students complete a longitudinal face from idea to realization, from basic working model to externalization and expression in the form of seminar and homework.
Teaching Content
Students shall select a daily space in Nanjing University under the instruction of teacher, according to personal interest and comprehensive requirement of content distribution, this space maybe indoor, single building or urban landscape. Students shall get familiar with relevant basic theory, classic case and historic fact of campus architecture in Nanjing University, and make a case analysis similar to "tracing". Conceptual design will first discuss object-oriented specific design tool and design methodology, then students shall make group according to similar or complementary daily space function, use similar or cooperative expression means for design drawing and modeling, and finally inspect design result in a set scene. Every design workgroup has four students who respectively complete a set of complete exercise, and cooperate to have course report.

生一年级
筑设计研究（一）：概念设计
鲁安东
程类型：必修
学分：40 学时／2 学分

duate Program 1st Year
SIGN STUDIO 1: CONCEPTUAL
SIGN · LU Andong
e: Required Course
dy Period and Credits: 40 hours/2 credits

课题内容
　　记忆·场所·叙事：一个宣言
研究主题
　　在我们这个时代，建筑学不应再是铺陈记忆的艺术，而是一种提醒的艺术。它为记忆的显影提供参照系，帮助人与场所之间建立起深沉的情感归属，进而得以触及人的灵魂。它是一种叙事的艺术。
　　记忆既是对当代建筑学的巨大挑战，也是其物质建造的终极命题。我们需要新的形式，叙事的形式，来涉足未知的记忆领域。
　　从记忆的角度来说，物质、活动、感动与存在之间不存在清晰的边界划分。它们共同构成了一次浮现，通过现在的再想象连接起过去与未来。
　　记忆的建筑学关注记忆在场所中的发生机制，而不是对记忆内容的选择性再现。记忆的建筑学关注主体参与的精神体验，而不是映射着环境的感知体验。它需要超越物质、空间与符号的新建筑语言，来沉浸、互动、质询与映射主体。正如现代主义背离了语言，用感知替代人与场所之间的意义关联，记忆的建筑学将重新回到语言的领域，并召唤叙事的心灵。
　　场所记忆的意义建构应该是开放的，而不是被给予的。设计和技术，应该帮助人找回自我与世界的归属感。它们必须走向存在的维度。这是当代建筑学的使命。
教学内容
　　此次"概念设计"课程选择了南京长江大桥作为研究对象。对南京长江大桥的记忆交织着宏大的历史记忆与亲密的个体记忆，曾是许许多多人建设、学习、旅行、工作、生活乃至想象的一部分。本课程从大桥记忆入手，探讨空间记忆的特征、发生机制和意义建构，最终设计一个当代的、公共的和富有创造性的记忆场所。

Subject Content
Memory · Place · Narrative: A Manifesto
Research Subject
At this time, architecture should not be an art to display memory, but an art of reminding. It provides reference to manifest memory, helps people and place establish profound emotional belonging, so as to touch the soul. It is an art of narration.
Memory is a huge challenge to contemporary architecture, and also the ultimate proposition of its material building. We need new form, form of narration to step in the unknown field of memory.
From the perspective of memory, substance, activity, touch and existence have no clear boundary. They compose an appearance, and connect the past and the future through current re-imagination.
Architecture of memory concerns the occurrence mechanism of memory in the place instead of selective reproduction of memory content. Architecture of memory concerns the spiritual experience of entity participation instead of reflecting the emotional experience of environment. It needs new architectural language that surpasses substance, space and symbol to immerse, interact, question and map the entity. Just as modernism deviated from language, perception replaces the meaning relation between people and place, architecture of memory will return to the field of language and recall the soul of narration.
The meaning construction of place memory should be open instead of being given. Design and technology should help people retrieve the sense of belonging of self and world. They must walk to the dimension of existence. This is the mission of contemporary architecture.
Teaching Content
In the course of conceptual design, Nanjing Yangtze River Bridge is selected as the research object. The memory of Nanjing Yangtze River Bridge interweaves grand history memory and close personal memory, and it is a part of construction, learning, travel, work, life and even imagination of many people. This course starts with the bridge memory, discusses on the characteristics, occurrence mechanism and meaning construction of space memory, and finally designs a contemporary, public and creative memory place.

研究生一年级

建筑设计研究（二）：建构设计
· 傅筱　孟宪川
课程类型：必修
学时学分：40 学时／2 学分

Graduate Program 1st Year

DESIGN STUDIO 2 : CONSTRUCTIONAL DESIGN · FU Xiao, MENG Xianchuan
Type: Required Course
Study Period and Credits: 40 hours/2 credits

课题内容

在江苏溧水无想山某场地内实地建造以竹结构为主的"山野乐园"景观小品，供游客使用。

教学目标

训练学生对设计概念与实地建造的关联性认知。

设计要点

设计研究分为两个阶段：一是设计阶段（2016.10.15—2017.1.15），二是建造阶段(2017年7月中旬，建造放在夏季比较适宜)。设计阶段需将图纸深化到施工图深度，学期结束将进行期终答辩，选出2~3组实施方案，并进一步优化设计。在最终的建造阶段，所有参与课程的同学都将前往工地参与实地建造，如人手短缺还将增招少年志愿者。

在设计研究过程中，根据经费状况将邀请国内知名学者、建筑师前来进行相关学术讲座和评图。

在设计建造过程中需配合业主单位进行记录和宣传工作。

Subject Content
Field construction of landscape sketch "Garden in Mountains" featuring bamboo structure on the site in Wuxi Mountain of Jiangsu, for the tourists.

Training Objective
Train the students to perceive the relevance between design concept and field construction.

Key Points of Design
The design study is divided into two phases: one is design phase (2016.10.15—2017.1.15), and the othe the construction phase (mid-July 2017, suitable for sum construction). Students need to deepen the drawings to phase of construction drawings. A final defense will be h and 2~3 buildable proposals will be chosen at the end of semester, the design will be further optimized. In the f implementation phase, all students attending this class will to the construction site and participate in the field construct In the case of shortage of manpower, we will also rec volunteers.

During the design and research, we will invite famous schol and architects in China to give relevant academic lectur attend the seminars and review the drawings.

During the design and construction process, it is necessary cooperate with the owner unit to record and publicize the wo

研究生一年级

建筑设计研究（二）：建构设计
· 陈浩如
课程类型：必修
学时学分：40 学时／2 学分

Graduate Program 1st Year

DESIGN STUDIO 2:CONSTRUCTIONAL DESIGN · CHEN Haoru
Type: Required Course
Study Period and Credits: 40 hours/2 credits

课题内容

在地自然建构工作营

教学目标

描述自然结构或建构。本次工作营的目的是通过在地自然建构的实践进行调查。将在八周内分为三个阶段进行。

教学内容

1. 在地描述

3 x Ms

测绘：等高线测绘、水体测绘、自然建筑测绘／人造建筑测绘、气候测绘。

建模：在地的物理和数字化表达。

材料：收集和编织的建构研究。

2. 田野营造

3. 自然建构

Subject Content
Insitu Natural-Tectonic Workshop

Training Objective
Description of natural construction, or tectonic. The aim of workshop is to investigate through the of insitu natural-tecto practice. It is set to be carried out in three phases during eight week period.

Teaching Content
1. Describing insitu

3 x Ms

Mapping: Contour mapping, water mapping, nature constr mapping/artificial construct mapping, climate mapping.
Modeling: physical and digital representation of the insitu.
Materials: collecting and woven tectonic studies.
2. Field building
3. Natural construction

研究生一年级
建筑设计研究（二）：城市设计
丁沃沃
课程类型：必修
学时学分：40学时／2学分

Graduate Program 1st Year
DESIGN STUDIO 2: URBAN DESIGN · DING Wowo
Type: Required Course
Study Period and Credits: 40 hours/2 credits

课题内容
　　城市更新与设计
教学目标
　　我国沿海发达城市城市化进程已经进入一个新的阶段，即由扩张型发展逐渐转为紧凑型发展，基于既有建成区的城市更新将成为城市建设的主要途径。在此过程中，通过改变土地使用性质、增大建设密度和改善交通组织等方式来提高土地使用效率，这些工作都是城市设计工作的主要任务，也将成为建筑设计工作的一项内容。本课题拟通过基于真实地块的设计实验，初步认知城市更新中面临的问题，了解城市建筑角色和城市物质空间的本质，掌握城市建筑与城市空间塑造之间的关系。此外，通过城市空间设计练习进一步深化空间设计的技能和方法。
关键术语
　　城市更新、立体交通、空间密度、城市形态
操作元素
　　"层""茎""界"与"空"
教学内容
　　1.以南京下关地块作为设计实验的场所，通过场地调研、场地地理条件认知、案例研习、设计分析和设计实验，探讨高效城市空间的建构方法和内涵。
　　2.通过设计拓展图示技能与表现方法，从空间意向出发建构城市物质空间的层与界面，由"空间意向""扫描"出城市空间的"层"与"界"。

Subject Content
Urban Renewal and Urban Design
Teaching Objective
The urbanization of developed coastal cities in China has stepped into a new stage and is gradually transforming from expanding development to compact development. Urban renewal based on existing built-up area will become a main route of urban construction. In this process, the main task of urban design work will mainly focus on change of land use, improvement of efficiency of land use and increase of building density, and become a part of architectural design work. Based on design experiment of real plot, this class aims to help the students cognize the problems in urban renewal, understand the nature of the roles of urban architecture and urban material space, and preliminarily master the relationship between unban architecture and urban space formation. In addition, the students will also practice the skills and methods of further deepening the space design through urban space design.
Key Terms
Urban renewal, three-dimensional transportation, spatial density, urban morphology
Operation Elements
"Level" "stem" "interface" and "space"
Teaching Content
1. This course takes Xiaguan block of Nanjing as the site for design experiment, to explore construction method and connotation of efficient urban space through site investigation, cognition of site geographical condition, case study, design analysis and design experiment.
2. This course will develop skills and methods of graphic expression through design, construct the level and interface of urban material space from space intention and "scan" the "level" and "interface" of urban space from "space intention".

研究生一年级
建筑设计研究（二）：城市设计
鲁安东
课程类型：必修
学时学分：40学时／2学分

Graduate Program 1st Year
DESIGN STUDIO 2 : URBAN DESIGN · LU Andong
Type: Required Course
Study Period and Credits: 40 hours/2 credits

课题内容
　　里下河聚落研究
教学内容
　　里下河地区的水乡聚落形态在20世纪水陆转换的过程中，发生着缓慢的变化。与江南水乡相比，其在20世纪的不稳定状态体现了新的社会、经济、交通、文化因素对聚落形态的影响。此次课程将以田野调查为基础，通过城市形态学的分析方法，对该地区的几个古镇进行设计研究。

Subject Content
Lixia River Settlement Research
Teaching Content
The form of waterside settlement in Lixia River area has slowly changed in the conversion of water and land in the 20th century. Compared to the water town in Yangtze River Delta, its unstable status in the 20th century reflects the influence of new society, economy, traffic, culture on the form of settlement. This course will base on field survey, through the analysis method of urban morphology, design and research several ancient towns in this area.

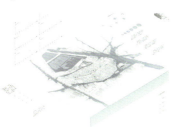

建筑理论课程
ARCHITECTURAL THEORY COURSES

本科二年级
建筑导论・赵辰等
课程类型：必修
学时/学分：36学时/2学分

Undergraduate Program 2nd Year
INTRODUCTION TO ARCHITECTURE • ZHAO Chen, etc.
Type: Required Course
Study Period and Credits:36 hours / 2 credits

课程内容
1. 建筑学的基本定义
   第一讲：建筑与设计/赵辰
   第二讲：建筑与城市/丁沃沃
   第三讲：建筑与生活/张雷
2. 建筑的基本构成
   （1）建筑的物质构成
   第四讲：建筑的物质环境/赵辰
   第五讲：建筑与节能技术/秦孟昊
   第六讲：建筑与生态环境/吴蔚
   第七讲：建筑与建造技术/冯金龙
   （2）建筑的文化构成
   第八讲：建筑与人文、艺术、审美/赵辰
   第九讲：建筑与环境景观/华晓宁
   第十讲：城市肌体/胡友培
   第十一讲：建筑与身体经验/鲁安东
   （3）建筑师职业与建筑学术
   第十二讲：建筑与表现/赵辰
   第十三讲：建筑与几何形态/周凌
   第十四讲：建筑与数字技术/吉国华
   第十五讲：城市与数字技术/童滋雨
   第十六讲：建筑师的职业技能与社会责任/傅筱

Course Content
I Preliminary of architecture
  1. Architecture and design / ZHAO Chen
  2. Architecture and urbanization / DING Wowo
  3. Architecture and life / ZHANG Lei
II Basic attribute of architecture
II-1 Physical attribute
  4. Physical environment of architecture / ZHAO Chen
  5. Architecture and energy saving / QIN Menghao
  6. Architecture and ecological environment / WU Wei
  7. Architecture and construction technology / FENG Jinlong
II-2 Cultural attribute
  8. Architecture and civilization, arts, aesthetic / ZHAO Chen
  9. Architecture and landscaping environment / HUA Xiaoning
  10. Urban tissue / HU Youpei
  11. Architecture and body / LU Andong
II-3 Architect: profession and academy
  12. Architecture and presentation / ZHAO Chen
  13. Architecture and geometrical form / ZHOU Ling
  14. Architectural and digital technology / JI Guohua
  15. Urban and digital technology / TONG Ziyu
  16. Architect's professional technique and responsibility / FU Xia

---

本科三年级
建筑设计基础原理・周凌
课程类型：必修
学时/学分：36学时/2学分

Undergraduate Program 3rd Year
BASIC THEORY OF ARCHITECTURAL DESIGN
• ZHOU Ling
Type: Required Course
Study Period and Credits:36 hours / 2 credits

教学目标
本课程是建筑学专业本科生的专业基础理论课程。本课程的任务主要是介绍建筑设计中形式与类型的基本原理。形式原理包含历史上各个时期的设计原则，类型原理讨论不同类型建筑的设计原理。
课程内容
1. 形式与类型概述
2. 古典建筑形式语言
3. 现代建筑形式语言
4. 当代建筑形式语言
5. 类型设计
6. 材料与建造
7. 技术与规范
8. 课程总结
课程要求
1. 讲授大纲的重点内容；
2. 通过分析实例启迪学生的思维，加深学生对有关理论及其应用、工程实例等内容的理解；
3. 通过对实例的讨论，引导学生运用所学的专业理论知识，分析、解决实际问题。

Training Objective
This course is a basic theory course for the undergraduate stude of architecture. The main purpose of this course is to introduce t basic principles of the form and type in architectural design. Fo theory contains design principles in various periods of history; ty theory discusses the design principles of different types of building
Course Content
1. Overview of forms and types
2. Classical architecture form language
3. Modern architecture form language
4. Contemporary architecture form language
5. Type design
6. Materials and construction
7. Technology and specification
8. Course summary
Course Requirement
1. Teach the key elements of the outline;
2. Enlighten students' thinking and enhance students' understandi of the theories, its applications and project examples throu analyzing examples;
3. Guide students using the professional knowledge to analysis a solve practical problems through the discussion of examples.

本科三年级
居住建筑设计与居住区规划原理・冷天 刘铨
课程类型：必修
学时/学分：36学时/2学分

Undergraduate Program 3rd Year
THEORY OF HOUSING DESIGN AND RESIDENTIAL PLANNING • LENG Tian, LIU Quan
Type: Required Course
Study Period and Credits:36 hours / 2 credits

课程内容
第一讲：课程概述
第二讲：居住建筑的演变
第三讲：套型空间的设计
第四讲：套型空间的组合与单体设计（一）
第五讲：套型空间的组合与单体设计（二）
第六讲：居住建筑的结构、设备与施工
第七讲：专题讲座：住宅的适应性，支撑体住宅
第八讲：城市规划理论概述
第九讲：现代居住区规划的发展历程
第十讲：居住区的空间组织
第十一讲：居住区的道路交通系统规划与设计
第十二讲：居住区的绿地景观系统规划与设计
第十三讲：居住区公共设施规划、竖向设计与管线综合
第十四讲：专题讲座：住宅产品开发
第十五讲：专题讲座：住宅产品设计实践
第十六讲：课程总结，考试答疑

Course Content
Lect. 1: Introduction of the course
Lect. 2: Development of residential building
Lect. 3: Design of dwelling space
Lect. 4: Dwelling space arrangement and residential building design (1
Lect. 5: Dwelling space arrangement and residential building design (2
Lect. 6: Structure, detail, facility and construction of residential building
Lect. 7: Adapt ability of residential building, supporting house
Lect. 8: Introduction of the theories of urban planning
Lect. 9: History of modern residential planning
Lect. 10: Organization of residential space
Lect. 11: Traffic system planning and design of residential area
Lect. 12: Landscape planning and design of residential area
Lect. 13: Public facilities and infrastructure system
Lect. 14: Real estate development
Lect. 15: The practice of residential planning and housing design
Lect. 16: Summary, question of the test

研究生一年级
现代建筑设计基础理论·张雷
课程类型：必修
学时/学分：18学时/1学分

Graduate Program 1st Year
PRELIMINARIES IN MODERN ARCHITECTURAL DESIGN • ZHANG Lei
Type: Required Course
Study Period and Credits: 18 hours/1 credit

课程内容
1. 现代设计思想的演变
2. 基本空间的组织
3. 建筑类型的抽象和还原
4. 材料运用与建造问题
5. 场所的形成及其意义
6. 今天的工作原则与策略

建筑可以被抽象到最基本的空间围合状态来面对它所必须解决的基本的适用问题，用最合理、最直接的空间组织和建造方式去解决问题，让普通材料和通用方法去回应复杂的使用要求，是建筑设计所应该关注的基本原则。

Course Content
1. Transition of the modern thoughts of design
2. Arrangement of basic space
3. Abstraction and reversion of architectural types
4. Material application and constructional issues
5. Formation and significance of sites
6. Nowaday working principles and strategies

Architecture can be abstracted to the most fundamental state of space enclosure, so as to confront all the basic applicable problems which must be resolved. The most reasonable and direct mode of space arrangement and construction shall be applied; ordinary materials and universal methods shall be used as the countermeasures to the complicated application requirement. These are the basic principles on which an architecture design institution shall focus.

研究生一年级
现代建筑设计方法论·丁沃沃
课程类型：必修
学时/学分：18学时/1学分

Graduate Program 1st Year
METHODOLOGY OF MODERN ARCHITECTURAL DESIGN • DING Wowo
Type: Required Course
Study Period and Credits: 18 hours/1 credit

课程内容
以建筑历史为主线，讨论建筑设计方法演变的动因/理念及其方法论。基于对传统中国建筑和西方古典建筑观念异同的分析，探索方法方面的差异。通过分析建筑形式语言的逻辑关系，讨论建筑形式语言的几何学意义。最后，基于城市形态和城市空间的语境探讨了建筑学自治的意义。

1. 引言
2. 西方建筑学的传统
3. 中国建筑的意义
4. 历史观与现代性
5. 现代建筑与意识的困境
6. 建筑形式语言的探索
7. 反思与回归理性
8. 结语

Course Content
Along the main line of architectural history, this course discussed the evolution of architectural design motivation/ideas and methodology. Due to different concepts between the Chinese architecture and Western architecture Matters. The way for analyzing and exploring has to be studied. By analyzing the logic relationship of architectural form language, the geometrical significance of architectural form language is explored. Finally, within the context of urban form and space, the significance of architectural autonomy has been discussed.

1. Introduction
2. Tradition of western architecture
3. Meaning of architecture in China
4. History and modernity
5. Modern architectural ideology and its dilemma
6. Exploration for architectural form language
7. Re-thinking and return to reason
8. Conclusion

研究生一年级
电影建筑学·鲁安东
课程类型：选修
学时/学分：36学时/2学分

Graduate Program 1st Year
CINEMATIC ARCHITECTURE • LU Andong
Type: Elective Course
Study Period and Credits: 36 hours/2 credits

课程内容
在本课程中，电影被视做一种独特的空间感知和思想交流的媒介。我们将学习如何使用电影媒介来对建筑和城市空间进行微观的分析和研究。本课程将通过循序渐进的教学帮助学生建立一种新的观察和理解建筑的方式，逐步培养学生的空间感知和空间想象的能力，以及使用电影媒介来交流自己的感觉和观点的能力。
本课程将综合课堂授课、实践操作和讲评讨论三种教学形式。在理论教学上，本课程将通过"普遍的运动""存在的直觉""组合的空间"和"城市的幻影"四讲逐步向学生介绍相关的历史和理论，特别是电影和空间的关系，并帮助学生理解相应的课堂练习。在实践操作教学上，本课程将通过四个小练习和三个大练习，帮助学生从镜头练习到空间表达到观点陈述逐步地掌握电影媒介并将其用于对自己设计能力的培养。而讲评讨论环节将向学生介绍电影媒介的技术和方法，并帮助学生对自己的动手经验进行反思。

Course Content
Film, in this course, is seen as a distinctive medium for the perception of space and the communication of thoughts. We shall learn how to use the unique narrative medium of film to conduct a microscopic study on architecture and urbanism. This course will teach the students of a new way of seeing and knowing architecture. Its purpose is not only to teach theories of urbanism and techniques of filmmaking, but also to teach the students, through a complete case study, of a cinematic (visceral and non-abstract) way of thinking, analyzing and presenting ideas.
This course is composed of a series of 4-hour sessions, which gradually lead the students to undertake their own research project and to produce a cinematic essay on a case study of their own choice. The teaching of this course will be conducted in three forms: the lectures will introduce the students to cinematic ways of seeing and understanding architecture; the tutorials will introduce the students to some basic cinematic techniques, including continuity editing, cinematography, storyboard, shooting script, and post-production; the seminars will review and discuss the students' works in several stages.

城市理论课程
URBAN THEORY COURSES

本科四年级
城市设计及其理论・丁沃沃 胡友培
课程类型：必修
学时/学分：36学时/2学分

Undergraduate Program 4th Year
THEORY OF URBAN DESIGN • DING Wowo, HU Youpei
Type: Required Course
Study Period and Credits: 36 hours / 2 credits

课程内容
第一讲 课程概述
第二讲 城市设计技术术语：城市规划相关术语；城市形态相关术语；城市交通相关术语；消防相关术语
第三讲 城市设计方法 —— 文本分析：城市设计上位规划；城市设计相关文献；文献分析方法
第四讲 城市设计方法 —— 数据分析：人口数据分析与配置；交通流量数据分析；功能分配数据分析；视线与高度数据分析；城市空间数据模型的建构
第五讲 城市设计方法 —— 城市肌理分类：城市肌理分类概述；肌理形态与建筑容量；肌理形态与开放空间；肌理形态与交通流量；城市绿地指标体系
第六讲 城市设计方法 —— 城市路网组织：城市道路结构与交通结构概述；城市路网功能；城市路网与城市空间；城市路网与市政设施；城市道路断面设计
第七讲 城市设计方法 —— 城市设计表现：城市设计分析图；城市设计概念表达；城市设计成果解析图；城市设计地块深化设计表达；城市设计空间表达
第八讲 城市设计的历史与理论：城市设计的历史意义；城市设计理论的内涵
第九讲 城市路网形态：路网形态的类型和结构；路网形态与肌理；路网形态的变迁
第十讲 城市空间：城市空间的类型；城市空间结构；城市空间形态；城市空间形态的变迁
第十一讲 城市形态学：英国学派；意大利学派；法国学派；空间句法
第十二讲 城市形态的物理环境：城市形态与物理环境；城市形态与环境研究；城市形态与环境测评；城市形态与环境操作
第十三讲 景观都市主义：景观都市主义的理论、操作和范例
第十四讲 城市自组织现象及其研究：城市自组织现象的魅力与问题；城市自组织系统研究方法；典型自组织现象案例研究
第十五讲 建筑学图式理论与方法：图式理论的研究；建筑学图式的概念；图式理论的应用；作为设计工具的图式；当代城市语境中的建筑学图式理论探索
第十六讲 课程总结

Course Content
Lect. 1. Introduction
Lect. 2. Technical terms: terms of urban planning, urba morphology, urban traffic and fire protection
Lect. 3. Urban design methods — documents analysis: urba planning and policies; relative documents; document analys techniques and skills
Lect. 4. Urban design methods — data analysis: data analysis demography, traffic flow, public facilities distribution, visual a building height; modelling urban spatial data
Lect. 5. Urban design methods — classification of urban fabric introduction of urban fabrics; urban fabrics and floor area rat urban fabrics and open space; urban fabrics and traffic flo criteria system of urban green space
Lect. 6. Urban design methods — organization of urban ro network: introduction; urban road network and urban functio urban road network and urban space; urban road network a civic facilities; design of urban road section
Lect. 7. Urban design methods — representation skills of U Design: mapping and analysis; conceptual diagram; analytic representation of urban design; representation of detail desig spatial representation of urban design
Lect. 8. Brief history and theories of urban design: historic meaning of urban design; connotation of urban design theories
Lect. 9. Form of urban road network: typology, structure a evolution of road network; road network and urban fabrics
Lect. 10. Urban space: typology, structure, morphology a evolution of urban space
Lect. 11. Urban morphology: Cozen School; Italian Schoo French School; Space Syntax Theory
Lect. 12. Physical environment of urban forms: urban form and physical environment; environmental study; environmen evaluation and environmental operations
Lect. 13. Landscape urbanism: ideas, theories, operations a examples of landscape urbanism
Lect. 14. Researches on the phenomena of the urban se organization: charms and problems; research methodology; cas studies of urban self-organization phenomena
Lect. 15. Theory and method of architectural diagram: theoretic study on diagrams; concepts of architectural diagrams; applicatio of diagram theory; diagrams as design tools; theoretical researc of architectural diagrams in contemporary urban context
Lect. 16. Summary

研究生一年级
城市形态研究・丁沃沃 赵辰 萧红颜
课程类型：必修
学时/学分：36学时/2学分

Graduate Program 1st Year
URBAN MORPHOLOGY • DING Wowo, ZHAO Chen, XIAO Hongyan
Type: Required Course
Study Period and Credits: 36 hours / 2 credits

课程要求
1. 要求学生基于对历史性城市形态的认知分析，加深对中西方城市理论与历史的理解。
2. 要求学生基于历史性城市地段的形态分析，提高对中西方城市空间特质及相关理论的认知能力。

课程内容
第一周 序言 概念、方法及成果
第二周 讲座1 城市形态认知的历史基础 —— 营造观念与技术传承
第三周 讲座2 城市形态认知的历史基础 —— 图文并置与意象构建
第四周 讲座3 城市形态认知的理论基础 —— 价值判断与空间生产
第五周 讲座4 城市形态认知的理论基础 —— 钩沉呈现与特征形塑
第六周 讲座5 历史城市的肌理研究
第七周 讲座6 整体与局部 —— 建筑与城市
第八周 讨论
第九周 讲座7 城市化与城市形态
第十周 讲座8 城市乌托邦
第十一周 讲座9 走出乌托邦
第十二周 讲座10 重新认识城市
第十三周 讲座11 城市设计背景
第十四周 讲座12 城市设计实践
第十五周 讲座13 城市设计理论
第十六周 讲评

Course Requirement
1. Deepen the understanding of Chinese and Western urba theories and histories based on the cognition and analysis historical urban form.
2. Improve the cognitive abilities of the characteristics a theories of Chinese and Western urban space based on th morphological analysis of the historical urban sites.
Course Content
Week 1. Preface — concepts, methods and results
Week 2. Lect. 1 Historical basis of urban form cognition - Developing concepts and passing of technologies
Week 3. Lect. 2 Historical basis of urban form cognition - Apposition of pictures and text and image construction
Week 4. Lect. 3 Theoretical basis of urban form cognition - Value judgment and space production
Week 5. Lect. 4 Theoretical basis of urban form cognition - History representation and feature shaping
Week 6. Lect. 5 Study on the grain of historical cities
Week 7. Lect. 6 Whole and part: Architecture and urban
Week 8. Discussion
Week 9. Lect. 7 Urbanization and urban form
Week 10. Lect. 8 Urban Utopia
Week 11. Lect. 9 Walk out of Utopia
Week 12. Lect. 10 Have a new look of the city
Week 13. Lect. 11 Background of urban design
Week 14. Lect. 12 Practice of urban design
Week 15. Lect. 13 Theory of urban design
Week 16. Discussions

本科四年级
景观规划设计及其理论 · 尹航
课程类型：选修
学时/学分：36学时/2学分

Undergraduate Program 4th Year
LANDSCAPE PALNNING DESIGN AND THEORY
YIN Hang
Type: Elective Course
Study Period and Credits: 36 hours / 2 credits

课程介绍
景观规划设计的对象包括所有的室外环境，景观与建筑的关系往往是紧密而且相影响的，这种关系在城市中尤为明显。景观规划设计及理论课程希望从景观设计理念、场地设计技术和建筑周边环境塑造等方面开展课程的教学，为建筑学本科生建立更加全面的景观知识体系，并且完善建筑学本科生在建筑场地设计、总平面规划与城市设计等方面的设计能力。

本课程主要从三个方面展开：一是理念与历史：以历史的视角介绍景观学科的发展过程，让学生对景观学科有一个宏观的了解，初步探索景观设计理念的发展；二是场地与文脉：通过阐述景观规划设计与周边自然环境、地理位置、历史文脉和方案可持续性的关系，建立场地与文脉的设计思维；三是景观与建筑：通过设计方法授课、先例分析作业等方式让学生增强建筑的环境意识，了解建筑的场地设计的影响因素、一般步骤与设计方法，并通过与"建筑设计六"和"建筑设计七"的设计任务书相配合的同步课程设计训练来加强学生景观规划设计的能力。

Course Description
The object of landscape planning design includes all outdoor environments; the relationship between landscape and building is often close and interactive, which is especially obvious in a city. This course expects to carry out teaching from perspective of landscape design concept, site design technology, building's peripheral environment creation, etc. to establish a more comprehensive landscape knowledge system for the undergraduate students of architecture, and perfect their design ability in building site design, master plane planning and urban design and so on.
This course includes three aspects:
1. Concept and history;
2. Site and context;
3. Landscape and building.

本科四年级
中西园林 · 许浩
课程类型：选修
学时/学分：36学时/2学分

Undergraduate Program 4th Year
GARDEN OF EAST AND WEST · XU Hao
Type: Elective Course
Study Period and Credits: 36 hours / 2 credits

课程介绍
帮助学生系统掌握园林、绿地的基本概念、理论和研究方法，尤其了解园林艺术的发展脉络，侧重各个流派如日式园林、江南私家园林、皇家园林、规则式园林、自由式园林、伊斯兰园林的不同特征和关系，使得学生能够从社会背景、环境等方面解读园林的发展特征，并能够开展一定的评价。

Course Description
Help students systematically master the basic concepts, theories and research methods of gardens and greenbelts, especially understand the evolution of gardening, emphasizing the different features and relationships of various genres, such as Japanese gardens, private gardens by the south of Yangtze River, royal gardens, rule-style gardens, free style gardens and Islamic gardens; enable students to interpret the characteristics of garden development from the point of view of social backgrounds, environment, etc. Furthermore to do the evaluation in depth.

研究生一年级
景观规划进展 · 许浩
课程类型：选修
学时/学分：18学时/1学分

Graduate Program 1st Year
LANDSCAPE PLANNING PROGRESS · XU Hao
Type: Elective Course
Study Period and Credits: 18 hours / 1 credit

课程介绍
生态规划是景观规划的核心内容之一。本课程总结了生态系统、生态保护和生态修复的基本概念。大规模生态保护的基本途径是国家公园体系，而生态修复则是通过人为干预对破损环境的恢复。本课程介绍了国家公园的价值、分类与成就，并通过具体案例论述了欧洲、澳洲景观设计过程中生态修复的做法。

Course Description
Ecological planning is one of the core contents of landscape planning. This course summarizes the basic concepts of ecological systems, ecological protection and ecological restoration. The basic channel of large-scale ecological protection is the national park system, while ecological restoration is to restore the damaged environment by means of human intervention. This course introduces the values, classification and achievements of national parks, and discusses the practices of ecological restoration in the process of landscape design in Europe and Australia.

研究生一年级
景观都市主义理论与方法 · 华晓宁
课程类型：选修
学时/学分：18学时/1学分

Graduate Program 1st Year
THEORY AND METHOD OF LANDSCAPE URBANISM
HUA Xiaoning
Type: Elective Course
Study Period and Credits: 18 hours / 1 credit

课程介绍
本课程介绍了景观都市主义思想产生的背景、缘起及其主要理论观点，并结合实例，重点分析了其在不同的场址和任务导向下发展起来的多样化的实践策略和操作性工具。通过这些内容的讲授，本课程的最终目的是拓宽学生的视野，引导学生改变既往的思维定式，以新的学科交叉整合的思路，分析和解决当代城市问题。

课程内容
第一讲　导论——当代城市与景观媒介
第二讲　生态过程与景观修复
第三讲　基础设施与景观嫁接
第四讲　嵌入与缝合
第五讲　水平性与都市表面
第六讲　城市图绘与图解
第七讲　AA景观都市主义——原型方法
第八讲　总结与作业

Course Description
The course introduces the backgrounds, the generation and the main theoretical opinions of landscape urbanism. With a series of instances, it particularly analyses the various practical strategies and operational techniques guided by various sites and projects. With all these contents, the aim of the course is to widen the students' field of vision, change their habitual thinking and suggest them to analyze and solve contemporary urban problems using the new ideas of the intersection and integration of different disciplines.

Course Content
Lect. 1: Introduction — contemporary cities and landscape medium
Lect. 2: Ecological process and landscape recovering
Lect. 3: Infrastructure and landscape engrafting
Lect. 4: Embedment and oversewing
Lect. 5: Horizontality and urban surface
Lect. 6: Urban mapping and diagram
Lect. 7: AA Landscape Urbanism — archetypical method
Lect. 8: Conclusion and assignment

历史理论课程
HISTORY THEORY COURSES

本科二年级
中国建筑史（古代）・赵辰 刘妍
课程类型：必修
学时/学分：36学时/2学分

Undergraduate Program 2nd Year
HISTORY OF CHINESE ARCHITECTURE (ANCIENT)
• ZHAO Chen, LIU Yan
Type: Required Course
Study Period and Credits:36 hours / 2 credits

教学目标
本课程作为本科建筑学专业的历史与理论课程，目标在于培养学生的史学研究素养与对中国建筑及其历史的认识两个层面。在史学理论上，引导学生理解建筑史学这一交叉学科的多种棱面与视角，并从多种相关学科层面对学生进行基本史学研究方法的训练与指导。中国建筑史层面，培养学生对中国传统建筑的营造特征与文化背景建立构架性的认识体系。

课程内容
中国建筑史学七讲与方法论专题。七讲总体走向从微观向宏观，整体以建筑单体-建筑群体-聚落与城市-历史地理为序；从物质到文化，建造技术-建造制度-建筑的日常性-纪念性-政治与宗教背景-美学追求。方法论专题包括建筑考古学、建筑技术史、人类学、美术史等层面。

Training Objective
As a mandatory historical & theoretical course for undergraduate students, this course aims at two aspects of training: the basic academic capability of historical research and the understanding of Chinese architectural history. It will help students to establish knowledge frame, that the discipline of History of Architecture as cross-discipline, is supported and enriched by multiple neighboring disciplines and that the features and development of Chinese Architecture roots deeply in the natural and cultural background.
Course Content
The course composes seven 4-hour lectures on Chinese Architecture and a series of lectures on methodology. The seven courses follow a route from individual to complex, from physical building to the intangible technique and to the cultural background, from technology to institution, to political and religious background and finally to aesthetic pursuit. The special topics on methodology include building archaeology, building science and technology, anthropology, art history and so on.

本科二年级
外国建筑史（古代）・胡恒
课程类型：必修
学时/学分：36学时/2学分

Undergraduate Program 2nd Year
HISTORY OF WESTERN ARCHITECTURE (ANCIENT)
• HU Heng
Type: Required Course
Study Period and Credits: 36 hours / 2 credits

教学目标
本课程力图对西方建筑史的脉络做一个整体勾画，使学生在掌握重要的建筑史知识点的同时，对西方建筑史在2000多年里的变迁的结构转folds（不同风格的演变）有深入的理解。本课程希望学生对建筑史的发展与人类文明发展之间的密切关联有所认识。

课程内容
1. 概论 2. 希腊建筑 3. 罗马建筑 4. 中世纪建筑
5. 意大利的中世纪建筑 6. 文艺复兴 7. 巴洛克
8. 美国城市 9. 北欧浪漫主义 10. 加泰罗尼亚建筑
11. 先锋派 12. 德意志制造联盟与包豪斯
13. 苏维埃的建筑与城市 14. 1960年代的建筑
15. 1970年代的建筑 16. 答疑

Training Objective
This course seeks to give an overall outline of Western architectural history, so that the students may have an in-depth understanding of the structural transition (different styles of evolution) of Western architectural history in the past 2000 years. This course hopes that students can understand the close association between the development of architectural history and the development of human civilization.
Course Content
1. Generality 2. Greek Architectures 3. Roman Architectures
4. The Middle Ages Architectures
5. The Middle Ages Architectures in Italy 6. Renaissance
7. Baroque 8. American Cities 9. Nordic Romanticism
10. Catalonian Architectures 11. Avant-Garde
12. German Manufacturing Alliance and Bauhaus
13. Soviet Architecture and Cities 14. 1960's Architectures
15. 1970's Architectures 16. Answer Questions

本科三年级
外国建筑史（当代）・胡恒
课程类型：必修
学时/学分：36学时/2学分

Undergraduate Program 3rd Year
HISTORY OF WESTERN ARCHITECTURE (MODERN)
• HU Heng
Type: Required Course
Study Period and Credits:36 hours / 2 credits

教学目标
本课程力图用专题的方式对文艺复兴时期的7位代表性的建筑师与位现当代的重要建筑师作品做一细致的讲解。本课程将重要建筑师的全部作品尽可能在课程中梳理一遍，使学生能够全面掌握重要建筑师的设计思想、理论主旨、与时代的特殊关联、在建筑史中的意义。

课程内容
1. 伯鲁乃列斯基 2. 阿尔伯蒂 3. 伯拉孟特
4. 米开朗基罗（1） 5. 米开朗琪罗（2） 6. 罗马诺
7. 桑索维诺 8. 帕拉蒂奥（1） 9. 帕拉蒂奥（2）
10. 赖特 11. 密斯 12. 勒・柯布西耶（1）
13. 勒・柯布西耶（2） 14. 海杜克 15. 妹岛和世
16. 答疑

Training Objective
This course seeks to make a detailed explanation to the works of 7 representative architects in the Renaissance period and important modern and contemporary architects in a special way. This course will try to reorganize all works of these important architects, so that the students can fully grasp their design ideas, theoretical subject and their particular relevance with the era and significance in the architectural history.
Course Content
1. Brunelleschi 2. Alberti 3. Bramante
4. Michelangelo(1) 5. Michelangelo(2)
6. Romano 7. Sansovino 8. Paratio(1) 9. Paratio(2)
10. Wright 11. Mies 12. Le Corbusier(1) 13. Le Corbusier(2)
14. Hejduk 15. Kazuyo Sejima
16. Answer Questions

本科三年级
中国建筑史（近现代）・赵辰
课程类型：必修
学时/学分：36学时/2学分

Undergraduate Program 3rd Year
HISTORY OF CHINESE ARCHITECTURE (MODERN)
• ZHAO Chen
Type: Required Course
Study Period and Credits:36 hours / 2 credits

课程介绍
本课程作为本科建筑学专业的历史与理论课程，是中国建筑史教学中的一部分。在中国与西方的古代建筑历史课程先学的基础上，了解中国社会进入近代，以至到现当代的发展进程。
在对比中西方建筑文化的基础之上，建立对中国近现代建筑的整体认识。深刻理解中国传统建筑文化在近代以来与西方建筑文化的冲突与相融之下，逐步演变发展至今天世界建筑文化的一部分之意义。

Course Description
As the history and theory course for undergraduate students of Architecture, this course is part of the teaching of History of Chinese Architecture. Based on the earlier studying of Chinese and Western history of ancient architecture, understand the evolution progress of Chinese society's entry into modern times and even contemporary age.
Based on the comparison of Chinese and Western building culture, establish the overall understanding of China's modern and contemporary buildings. Have further understanding of the significance of China's traditional building culture's gradual evolution into one part of today's world building culture under conflict and blending with Western building culture in modern times.

研究生一年级
建筑理论研究 · 王骏阳
课程类型：必修
学时/学分：18学时/1学分

Graduate Program 1st Year
STUDY OF ARCHITECTURAL THEORY · WANG Junyang
Type: Required Course
Study Period and Credits: 18 hours / 1 credit

课程介绍
本课程是西方建筑史研究生教学的一部分。主要涉及当代西方建筑界具有代表性的思想和理论，其主题包括历史主义、先锋建筑、批判理论、建构文化以及对当代城市的解读等。本课程大量运用图片资料，广泛涉及哲学、历史、艺术等领域，力求在西方文化发展的背景中呈现建筑思想和理论的相对独立性及关联性，理解建筑作为一种人类活动所具有的社会和文化意义，启发学生的理论思维和批判精神。

课程内容
第一讲 建筑理论概论
第二讲 建筑自治
第三讲 柯林·罗：理想别墅的数学与其他
第四讲 阿道夫·路斯与装饰美学
第五讲 库哈斯与当代城市的解读
第六讲 意识的困境：对现代建筑的反思
第七讲 弗兰普顿的建构文化研究
第八讲 现象学

Course Description
This course is a part of teaching Western architectural history for graduate students. It mainly deals with the representative thoughts and theories in Western architectural circles, including historicism, vanguard building, critical theory, construction culture and interpretation of contemporary cities and more. Using a lot of pictures involving extensive fields including philosophy, history, art, etc., this course attempts to show the relative independence and relevance of architectural thoughts and theories under the development background of Western culture, understand the social and cultural significance owned by architectures as human activities, and inspire students' theoretical thinking and critical spirit.

Course Content
Lect. 1. Introduction to architectural theories
Lect. 2. Autonomous architecture
Lect. 3. Colin Rowe : the mathematics of the ideal villa and others
Lect. 4. Adolf Loos and adornment aesthetics
Lect. 5. Koolhaas and the interpretation of contemporary cities
Lect. 6. Conscious dilemma: the reflection of modern architecture
Lect. 7. Studies in tectonic culture of Frampton
Lect. 8. Phenomenology

研究生一年级
建筑史研究 · 胡恒
课程类型：选修
学时/学分：36学时/2学分

Graduate Program 1st Year
ARCHITECTURAL HISTORY RESEARCH · HU Heng
Type: Elective Course
Study Period and Credits: 36 hours / 2 credits

教学目标
本课程的目的有二。其一，通过对建筑史研究的方法做一概述，来使学生粗略了解西方建筑史研究方法的总的状况。其二，通过对当代史概念的提出，且用若干具体的案例研究，来向学生展示当代史研究的路数、角度、概念定义、结构布置、主题设定等内容。

课程内容
1. 建筑史方法概述（1）
2. 建筑史方法概述（2）
3. 建筑史方法概述（3）
4. 塔夫里的建筑史研究方法
5. 当代史研究方法 —— 周期
6. 当代史研究方法 —— 杂交
7. 当代史研究方法 —— 阈限
8. 当代史研究方法 —— 对立

Training Objective
This course has two objectives: 1. Give the students a rough understanding of the overall status of the research approaches of the Western architectural history through an overview of them. 2. Show students the approaches, point of view, concept definition, structure layout, theme settings and so on of contemporary history study through proposing the concept of contemporary history and several case studies.

Course Content
1. The overview of the method of architectural history(1)
2. The overview of the method of architectural history(2)
3. The overview of the method of architectural history(3)
4. Tafuri's study method of architectural history
5. The study method of contemporary history — period
6. The study method of contemporary history — hybridization
7. The study method of contemporary history — limen
8. The study method of contemporary history — opposition

建筑技术课程
ARCHITECTURAL TECHNOLOGY COURSES

本科二年级
CAAD理论与实践·童滋雨
课程类型：必修
学时/学分：36学时/2学分

Undergraduate Program 2nd Year
THEORY AND PRACTICE OF CAAD · TONG Ziyu
Type: Required Course
Study Period and Credits: 36 hours / 2 credits

课程介绍
在现阶段的CAD教学中，强调了建筑设计在建筑学教学中的主干地位，将计算机技术定位于绘图工具，本课程就是帮助学生可以尽快并且熟练地掌握如何利用计算机工具进行建筑设计的表达。课程中整合了CAD知识、建筑制图知识以及建筑表现知识，将传统CAD教学中教会学生用计算机绘图的模式向教会学生用计算机绘制有形式感的建筑图的模式转变，强调准确性和表现力作为评价CAD学习的两个最重要指标。
本课程的具体学习内容包括：
    1. 初步掌握AutoCAD软件和SketchUP软件的使用，能够熟练完成二维制图和三维建模的操作；
    2. 掌握建筑制图的相关知识，包括建筑投影的基本概念、平立剖面、轴测、透视和阴影的制图方法和技巧；
    3. 图面效果表达的技巧，包括黑白线条图和彩色图纸的表达方法和排版方法。

Course Description
The core position of architectural design is emphasized in the CA course. The computer technology is defined as drawing instrume The course helps students learn how to make architectu presentation using computer fast and expertly. The knowled of CAD, architectural drawing and architectural presentation a integrated into the course. The traditional mode of teaching stude to draw in CAD course will be transformed into teaching stude to draw architectural drawing with sense of form. The precision a expression will be emphasized as two most important factors estimate the teaching effect of CAD course.
Contents of the course include:
1. Use AutoCAD and SketchUP to achieve the 2-D drawing and 3 modeling expertly.
2. Learn relational knowledge of architectural drawing, includ basic concepts of architectural projection, drawing methods a skills of plan, elevation, section, axonometry, perspective a shadow.
3. Skills of presentation, including the methods of expression a lay out using mono and colorful drawings

---

本科三年级
建筑技术1——结构、构造与施工·傅筱
课程类型：必修
学时/学分：36学时/2学分

Undergraduate Program 3rd Year
ARCHITECTURAL TECHNOLOGY 1 — STRUCTURE, CONSTRUCTION AND EXECUTION · FU Xiao
Type: Required Course
Study Period and Credits:36 hours / 2 credits

课程介绍
本课程是建筑学专业本科生的专业主干课程。本课程的任务主要是以建筑师的工作性质为基础，讨论一个建筑生成过程中最基本的三大技术支撑（结构、构造、施工）的原理性知识要点，以及它们在建筑实践中的相互关系。

Course Description
The course is a major course for the undergraduate students architecture. The main purpose of this course is based on the natu of the architect's work, to discuss the principle knowledge poir of the basic three technical supports in the process of generati construction (structure, construction, execution), and their mutu relations in the architectural practice.

本科三年级
建筑技术2——建筑物理·吴蔚
课程类型：必修
学时/学分：36学时/2学分

Undergraduate Program 3rd Year
ARCHITECTURAL TECHNOLOGY 2 — BUILDING PHYSICS · WU Wei
Type: Required Course
Study Period and Credits:36 hours / 2 credits

课程介绍
本课程是针对三年级学生所设计，课程介绍了建筑热工学、建筑光学、建筑声学中的基本概念和基本原理，使学生能掌握建筑的热环境、声环境、光环境的基本评估方法，以及相关的国家标准。完成学业后在此方向上能阅读相关书籍，具备在数字技术方法等相关资料的帮助下，完成一定的建筑节能设计的能力。

Course Description
Designed for the Grade-3 students, this course introduces the ba concepts and basic principles in architectural thermal engineeri architectural optics and architectural acoustics, so that the stude can master the basic methods for the assessment of buildin thermal environment, sound environment and light environment well as the related national standards. After graduation, the stude will be able to read the related books regarding these aspec and have the ability to complete certain building energy efficien designs with the help of the related digital techniques and method

---

本科三年级
建筑技术3——建筑设备·吴蔚
课程类型：必修
学时/学分：36学时/2学分

Undergraduate 3rd Year
ARCHITECTURAL TECHNOLOGY 3 — BUILDING EQUIPMENT · WU Wei
Type: Required Course
Study Period and Credits:36 hours / 2 credits

课程介绍
本课程是针对南京大学建筑与城市规划学院本科学生三年级所设计。课程介绍了建筑给水排水系统、采暖通风与空气调节系统、电气工程的基本理论、基本知识和基本技能，使学生能熟练地阅读水电、暖通工程图，熟悉水电与消防的设计、施工规范，了解燃气供应、安全用电及建筑防火、防雷的初步知识。

Course Description
This course is an undergraduate class offered in the School Architecture and Urban Planning, Nanjing University. It introduc the basic principle of the building services systems, the techniq of integration amongst the building services and the buildin Throughout the course, the fundamental importance to ener ventilation, air-conditioning and comfort in buildings are highlighte

---

研究生一年级
传热学与计算流体力学基础·郜志
课程类型：选修
学时/学分：18学时/1学分

Graduate Program 1st Year
FUNDAMENTALS OF HEAT TRANSFER AND COMPUTATIONAL FLUID DYNAMICS · GAO Zhi
Type: Elective Course
Study Period and Credits: 18 hours / 1 credit

课程介绍
本课程的主要任务是使建筑学/建筑技术学专业的学生掌握传热学和计算流体力学的基本概念和基础知识，通过课程教学，使学生熟悉传热学中导热、对流和辐射的经典理论，了解传热学和计算流体力学的实际应用和最新研究进展，为建筑能源与环境系统的计算和模拟打下坚实的理论基础。教学中尽量简化传热学和计算流体力学经典课程中复杂公式的推导过程，而着重于如何解决建筑能源与建筑环境中涉及流体流动和传热的实际应用问题。

Course Description
This course introduces students majoring in building science a engineering / building technology to the fundamentals of he transfer and computational fluid dynamics (CFD). Students w study classical theories of conduction, convection and radiati heat transfers, and learn advanced research developments heat transfer and CFD. The complex mathematics and physi equations are not emphasized. It is desirable that for real-ca scenarios students will have the ability to analyze flow and he transfer phenomena in building energy and environment systems

研究生一年级
建筑节能与可持续发展 · 秦孟昊
课程类型: 选修
学时/学分: 18学时/1学分

Graduate Program 1st Year
ENERGY CONSERVATION AND SUSTAINABLE ARCHITECTURE · QIN Menghao
Type: Elective Course
Study Period and Credits: 18 hours / 1 credit

课程介绍

随着我国建筑总量的攀升和居住舒适度的提高，建筑能耗急剧上升，建筑节能成为影响能源安全和提高能效的重要因素之一。建筑节能的关键首先是要设计"本身节能的建筑"，建筑师必须从建筑设计的最初阶段，在建筑物的形体、结构、开窗方式、外墙选材等方面融入节能设计的定量分析。而这些很难通过传统建筑设计方法达到，必须依靠建筑技术、建筑设备等多学科互动协作方才能完成。这已成为世界各大建筑与城市规划学院教学的一个重点。

本课程将采用双语教学，主要面向建筑设计专业学生讲授建筑物理、建筑技术专业关于建筑节能方面的基本理念、设计方法和模拟软件，并指导学生将这些知识互动运用到节能建筑设计的过程中，在建筑设计专业和建筑技术专业之间建立一个互动的平台，从而达到设计"绿色建筑"的目标，并为以后开展交叉学科研究、培养复合型人才奠定基础。

Course Description

With the rising of China's total number of buildings and the need for living comfort, building energy consumption is rising sharply. Building energy efficiency has become one of the key factors influencing the energy security and energy efficiency. The first key for building energy efficiency is to design "a building that conserves energy itself" and architects must carry out planning at the very beginning of building design. However, it is difficult to satisfy them by means of traditional architectural design approaches; it must be realized by interactive collaboration of diversified subjects including construction technology, construction equipment, etc. Strengthening the interaction of architectural design specialties and construction technology specialties in designing has become a key point in this course as well as in the teaching of various large architecture and urban planning colleges around the world.

研究生一年级
建筑环境学 · 郜志
课程类型: 选修
学时/学分: 18学时/1学分

Graduate Program 1st Year
FUNDAMENTALS OF BUILT ENVIRONMENT · GAO Zhi
Type: Elective Course
Study Period and Credits: 18 hours / 1 credit

课程介绍

本课程的主要任务是使建筑学/建筑技术学专业的学生掌握建筑环境的基本概念，学习建筑与城市热湿环境、风环境和空气质量的基础知识。通过课程教学，使学生熟悉城市微气候等理论，并了解人体对热湿环境的反应，掌握建筑环境学的实际应用和最新研究进展，为建筑能源和环境系统的测量与模拟打下坚实的基础。

Course Description

This course introduces students majoring in building science and engineering / building technology to the fundamentals of built environment. Students will study classical theories of built / urban thermal and humid environment, wind environment and air quality. Students will also familiarize urban micro environment and human reactions to thermal and humid environment. It is desirable that students will have the ability to measure and simulate building energy and environment systems based upon the knowledge of the latest development of the study of built environment.

研究生一年级
材料与建造 · 冯金龙
课程类型: 必修
学时/学分: 18学时/1学分

Graduate Program 1st Year
MATERIAL AND CONSTRUCTION · FENG Jinlong
Type: Required Course
Study Period and Credits: 18 hours / 1 credit

课程介绍

本课程将介绍现代建筑技术的发展过程，论述现代建筑技术及其美学观念对建筑设计的重要作用。探讨由材料、结构和构造方式所形成的建筑建造的逻辑方式研究。研究建筑形式产生的物质技术基础，诠释现代建筑的建构理论与研究方法。

Course Description

It introduces the development process of modern architecture technology and discusses the important role played by the modern architecture technology and its aesthetic concepts in the architectural design. It explores the logical methods of construction of the architecture formed by materials, structure and construction. It studies the material and technical basis for the creation of architectural form, and interprets construction theory and research methods for modern architectures.

研究生一年级
计算机辅助技术 · 吉国华
课程类型: 选修
学时/学分: 36学时/2学分

Graduate Program 1st Year
TECHNOLOGY OF CAAD · JI Guohua
Type: Elective Course
Study Period and Credits: 36 hours / 2 credits

课程介绍

随着计算机辅助建筑设计技术的快速发展，当前数字技术在建筑设计中的角色逐渐从辅助绘图转向了真正的辅助设计，并引发了设计的革命和建筑的形式创新。本课程讲授AutoCAD VBA和RhinoScript编程。让学生在掌握"宏"/"脚本"编程的同时，增强以理性的过程思维方式分析和解决设计问题的能力，为数字建筑设计打下必要的基础。

课程分为三个部分:
1. VB语言基础，包括VB基本语法、结构化程序、数组、过程等编程知识和技巧；
2. AutoCAD VBA，包括AutoCAD VBA的结构、二维图形、人机交互、三维对象等，以及基本的图形学知识；
3. RhinoScript概要，包括基本概念、Nurbs概念、VBScript简介、曲线对象、曲面对象等。

Course Description

Following its fast development, the role of digital technology in architecture is changing from computer-aided drawing to real computer-aided design, leading to a revolution of design and the innovation of architectural form. Teaching the programming with AutoCAD VBA and RhinoScript, the lecture attempts to enhance the students' capability of reasoningly analyzing and solving design problems other than the skills of "macro" or "script" programming, to let them lay the base of digital architectural design.
The course consists of three parts:
1. Introduction to VB, including the basic grammar of VB, structural program, array, process, etc.
2. AutoCAD VBA, including the structure of AutoCAD VBA, 2D graphics, interactive methods, 3D objects, and some basic knowledge of computer graphics.
3. Brief introduction of RhinoScript, including basic concepts, the concept of Nurbs, summary of VBScript, and Rhino objects.

研究生一年级
GIS基础与应用 · 童滋雨
课程类型: 选修
学时/学分: 18学时/1学分

Graduate Program 1st Year
CONCEPT AND APPLICATION OF GIS · TONG Ziyu
Type: Elective Course
Study Period and Credits: 18 hours / 1 credit

课程介绍

本课程的主要目的是让学生理解GIS的相关概念以及GIS对城市研究的意义，并能够利用GIS软件对城市进行分析和研究。

Course Description

This course aims to enable students to understand the related concepts of GIS and the significance of GIS to urban research, and to be able to use GIS software to carry out urban analysis and research.

回声——来自毕业生的实践
ECHO—FROM PRACTICES OF GRADUATES

建筑设计研究（二） DESIGN STUDIO 2

我的建造教学与南大建筑的渊源
MY TEACHING OF CONSTRUCTION AND ORIGIN WITH ARCHITECTURE OF NJU

钟冠球

我从2009年开始在华南理工建筑学院任教，除了设计课以外，还担任一门低年级的模型建造课的主讲老师，在课中，让学生用真实的材料尝试去建构一些经典案例的原型，这也逐渐引起学生的兴趣，学生一届比一届卖力。同时，我参与到学院模型实验室的建设中，购置了多台激光切割机、CNC铣床和木工设备，并单独配置一个操作间，制定使用制度，将模型室开放给全学院的师生使用。在模型室硬件的支持下，学院能够在建造训练上大胆迈步，实现结构、构造、材料教学的更好结合。

2015—2017年，我作为华南理工大学营造竞赛的出题人，组织了三年三次建造竞赛。营造竞赛是华南理工大学建筑学院的一个课外的传统竞赛，从制作最小的灯具到建造尺度越来越大的构筑物，至今已近20年。之前的营造竞赛没有严格的题目限定，队伍之间没法精细化比较，并且基本是院内竞赛。我认为竞赛规模化、正规化、竞争性是非常重要的，因此引入了赞助商支持，增加竞赛获奖奖金和建造补贴，增加了更严格的初赛环节，邀请外校队伍参赛，由学生会各部门组成竞赛组委会，加大对外宣传和增加竞赛的重要程度……一个好玩又刺激的挑战，没有学生不爱。

2015年题目是"模块化木构展览小筑集群设计"，规定2.4m×3.6m×2.4m（高）尺寸的小建筑，十支入选决赛的小筑不但各自独立地存在，呈现自身的特色，而且共同成为一个整体，形成一个串联式的展廊。在"模块化"要求下，搭建作品从结构到围护甚至螺丝钉的位置都要遵循一定的逻辑关系，需要小组在场地以外预制好各部件，在规定的较短时间内进行快速安装。

2016年题目是"共享木构设施"，尺寸更大了，2.4m×4.8m×3.6m（高），要求学生制作一个可供学习、生活、活动用的小型木屋。学生从自身使用需求出发，创造了很多可变的多功能的小房子。在建成后日常中非常多的学生使用这些临时设施，为建筑学院增加了100m²临时使用面积。竞赛后来设立了"维护运营奖"，在竞赛一年后进行评比，看哪些作品能很好地经受时间和气候的检验，哪些作品能被人更好地使用下去，参与决赛的同学开始获得了"使用后评价"的思考方式。

2017年题目是"光影竹亭"，尺寸限定为3m×3m×4.2m以下，顾名思义，换成了竹子作为主材，竹子的成本比木构要低很多，但是加工难度更大，特别是竹子的热弯。因校内明火操作比较危险，为此，组委会在校外借用了临时场地，由参加过竹子建构的

指导小组教会参赛同学如何热弯竹子，然后由学生自己完成全部的操作。预制加工成的竹子运回校内进行拼装。学生营造作品为校园创造了很多特色空间，吸引了很多专业和非专业的老师和同学过来观看和留影。

2015—2017年三年的竞赛也吸引了深圳大学、广州美术学院、广州大学、广东工业大学等多所学校参与，逐渐演变成华南地区一个影响力较大的建造竞赛。而作为该年的组织者，有很多思路都是来源于我在南大建筑读研究生时亲身参与的建造课程。

2005年的时候，我在南京大学研究生的建造课程，由赵辰老师、冯金龙老师和丁凌老师主持，以2.4m×2.4m×2.4m的木构框架作为基本单元，学生不但要自己搭起1:1的木构架，还要设计围护结构并用真实材料建造出来，这给学生已有的构造知带来巨大的挑战。搭建在一个地下室里进行，搭建成果在当时对国内高校起到很好的示范作用。

引用南京大学赵辰老师对2005年木构建造的回忆来描绘建造之后的收获再恰不过了："我们还搞了个所谓评图，不过说实话，能得到什么样的讲评，我根本不兴趣……无论哪个评委怎么讲都无所谓，你们都要很快乐，然后接受它。盖房子该是这样，你已经盖出来了，就不用管别人怎么说了，对你们来说评图应该是一个典。"

实际上，设计—深化—采购—加工—组装—建成，这一环环相扣的建造实验当与真正的建筑生产过程是相似的，是微缩化实现的建筑过程。这个过程中，学生不仅仅"纸上谈兵"地设计，而是因循结构和构造的逻辑，体验手工操作和机器生产及预制装配的过程。

建造教学提供了很好的感受重力和材料的机会，亲身建造者经常会遇到高空作业的难题，而这些都是通过虚拟模型无法体会的。建造教育使学生同时获得对结构、造、材料的了解，应鼓励他们探索、研究非常规节点、非成熟做法。既然是实验，要允许犯错，允许失败。包豪斯早在1919年，就将"制作"训练列入建筑教育有计训练内容具体包括石刻、木刻、陶艺、金属加工、木工、纺织、浇铸等，每位学生要到工坊里进行学习，以熟知加工工艺，更重要的是对材料的属性有了直接的认相较于以视觉、感知、构形为先导的布扎体系，包豪斯以材料试验和材料操作作

础课程,再演化成建造上更专业的建造结构、外立面设计、采暖通风设计、照明设
力学计算、概预算等课程。

就国内而言,台湾可能比大陆更早在建筑教育体系中开展建造教学,较早有淡江
学建筑系黄瑞茂先生在淡水地区的社区营造,学生深入社区寻找需要建造的点,参
设计和完善社区配套功能房间的搭建,还有参加社区艺术庆典活动的装置设计和建
后来辐射到大陆的有所谓"在野建筑学教育"的谢英俊主持的乡村建筑工作室,
在晏阳建设的地球村,召集国内很多建筑学子参与建造工作坊,这种身体力行的建
训练,是建筑教育体系必要的补充。2008年坂茂在四川的"纸屋"华林小学校舍建
也类似。参加了这些建造活动的学子获得了很多珍贵的经历,现已成为一批更关注
建造的青年建筑师群体。

近些年来,不少毕业于南大建筑的建筑师做出了一系列细节很好的建筑作品,我
这些与最初南大建筑的建造课程有关,南大建筑学生普遍关注节点构造,"三句不
细部",希望这些能够让南大建筑继续独树一帜。

ave taught in School of Architecture of South China University of Technology since
09, besides the design course, I am also the chief lecturer of modeling course of
nior grade. In the class, I make students use real material to try to construct the
ototype of some classic cases, and this gradually arouses the interest of students,
o make more efforts than their predecessors. At the same time, I have participated
the construction of model laboratory of the school, purchased several lasercuts,
NC millers and woodworking equipment, configured a separate operation room,
veloped use system, and opened the model room to all teachers and students
the school. Under the hardware support of model room, the school can boldly
ake progress in construction training, and realize better combination of structure,
nstruction, material teaching.

From 2015 to 2017, as the question maker of construction competition of South China University of Technology, I organized three construction competitions in three years. Construction competition is a traditional extracurricular competition of South China University of Technology, and it has a history of nearly 20 years from making the smallest lamp to structure with greater and greater size. The construction competition in the past had no strict title limit, there was no fine comparison between teams, and it was basically competition in the university. I think that the scale, regularization and competitiveness of competition are very important, therefore, I introduce sponsorship, increase competition bonus and construction subsidy, increase stricter preliminary competition, invite teams out of the university, make departments of student union establish competition organizing committee, strengthen external propaganda and increase the importance of the competition…A funny and exciting challenge, every student loves it.

Title in 2015 was "Modular wood construction exhibition small building cluster design", which regulated small building of 2.4m×3.6m×2.4m (height), ten small buildings selected to the final not only respectively presented their own characteristics, but also composed a whole and formed an exhibition gallery in series. Under the requirement of "modularization", certain logic relation shall be observed from structure to fencing and even the position of screw, each team shall prefabricate all parts out of the site, and quickly install within the regulated time.

Title in 2016 was "Share wood construction facility", with greater size, 2.4m×4.8m×3.6m (height), students were required to make a small wood cabin for learning, living and activity. Students started with their own demand, created many variable and multifunctional small houses. After completion, many students used

these temporary facilities, and additional 100m² temporary use area was increased for the School of Architecture. Later the competition established a "maintenance operation award", which was selected one year after the competition, checking which work can well withstand the test of time and weather, which work can be better used, and the students that participate in the final obtain the thinking mode of "evaluating after use".

Title in 2017 is "Light and shadow bamboo dome", the size is limited to 3m×3m×4.2m below, as the name implies, bamboo becomes the main material, although the cost of bamboo is lower than wood, the processing difficulty is greater, especially the hot bending of bamboo. However, it is dangerous to have open fire operation in the university, therefore, the organizing committee borrows temporary site out of the university, the guiding team that has participated in bamboo construction teaches the students how to hot bend bamboo, and then students independently complete all operations. The prefabricated bamboo is transported back to the university for assembly. Works created by students create many special space for the campus, and attract many professional and non-professional teachers and students to view and take picture.

The competition from 2015 to 2017 also attracts a number of schools including Shenzhen University, Guangzhou Academy of Fine Arts, Guangzhou University, Guangdong University of Technology, etc., and gradually becomes an influential construction competition in South China. As the organizer in these three years, I have many ideas from my personally experienced construction course when I was a postgraduate in School of Architecture of Nanjing University.

In 2005, I was a postgraduate of construction in Nanjing University, lectured by teacher ZHAO Chen, teacher FENG Jinlong and teacher ZHOU Ling. 2.4m×2.4m×2.4m wood frame was used as the basic unit, students did not only independently build 1:1 wood structure, but also designed the fencing structure a constructed by real material, which was a huge challenge to the existing construct knowledge of students. The building was conducted in a basement, and the build achievement played a good demonstration role to domestic colleges at that momer It is appropriate to quote the memory of teacher ZHAO Chen from Nanjing Univers on the wood construction in 2005 to describe the harvest after construction: "we a have a so-called drawing review, to be honest, I am not interested in the comment all…it does not matter which judge makes comment, you should be happy and th accept it. Building a house should be like this, you have built it, do not care w others say about it, drawing review should be a celebration to you."

Actually, design—detailing-purchase-processing-assembly-completion, t interconnected construction test process is similar to the real building producti process, and it is the miniature of construction process. In this process, studer not only theorize the design, but also follow the logic of structure and constructi experience the process of manual operation and machine production a prefabrication assembly.

Construction teaching provides good opportunity to feel gravity and material, t difficulty in altitude operation that the constructor personally experiences cann be felt through virtual model. Construction education makes students understa structure, construction, material, and encourages them to explore, research irregu node, and immature method. Since it's an experiment, it allows mistake and failu Bauhaus has listed "making" training in the architectural education plan early 1919, and the training content includes stone carving, wood carving, ceramic a metal processing, woodworking, textile, casting, etc., every student shall study the workshop to well know processing technic, moreover, direct recognition of t attribute of material. Compared to Beaux-Arts system guided by vision, perceptic

composition, Bauhaus takes material test and material operation as the fundamental course, then evolves to more professional courses as construction structure, facade design, heating and ventilation design, lighting design, mechanical calculation, budget estimate, etc. in construction.

Taiwan may be earlier than Chinese mainland to start construction teaching in architectural education system. Mr. HUANG Ruimao from Department of Architecture of Tamkang University built community in Tamsui, students went deep into the community and sought construction site, participated in designing and improving the building of community supporting functional room, and participated in device design and construction of community art festival. Later, it radiates to the countryside architecture studio hosted by XIE Yingjun who advocates "Architecture education in the wild" in Chinese mainland. He constructs earth village in Yanyang, and convenes many domestic architecture students to participate in the construction, and this personally practiced construction training is the necessary supplementation of architecture education system. In 2008, the "paper house" Hualin Primary School constructed by Shigeru BAN in Sichuan is also the similar case. Students that participated in these construction activities obtained precious experience, and now have become a young architect group that attaches more importance to construction. In recent years, many architects graduated from School of Architecture of Nanjing University have completed a series of architectural works with good details, and I think these are related to the original construction course of School of Architecture of Nanjing University. Students of School of Architecture of Nanjing University generally concern node construction, "everything is related to detail", and I hope that these can make School of Architecture of Nanjing University continue to develop a school of its own.

在建成后日常中非常多的学生都使用这些临时设施，为建筑学院增加了100 m² 临时使用面积，竞赛后来设立了"维护运营奖"，在竞赛一年后进行评选，看哪些作品能很好地经受时间和气候的检验，哪些作品能被人更好地使用下去，参与竞赛的同学开始获得了"使用后评价"的思考方式。
After completion, many students used these temporary facilities, and additional 100 m² temporary used area was increased for the School of Architecture. Later the competition established a "maintenance operation award", which was selected one year after the competition, checking which work can well withstand the test of time and weather, which work can be better used, and the students that participate in the final obtained the thinking mode of "evaluating after use".

2017年题目是"光影竹穹"。学生营造作品为校园创造了很多特色空间，吸引了很多专业和非专业的老师和同学过来观看和留影。Title in 2017 is "Light and shadow bamboo dome". Works created by students create many special space for the campus, and attract many professional and non-professional teachers and students to view and take picture.

其他
MISCELLANEA

讲座
Lectures

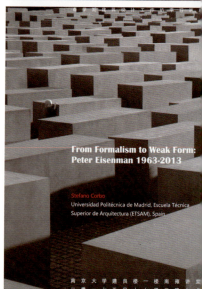

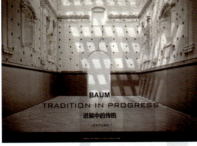
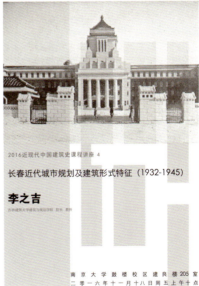

硕士学位论文列表
List of Thesis for Master Degree

研究生姓名	研究生论文标题	导师姓名
高 翔	当代实体书店转型设计研究	张 雷
刘文沛	当代购物中心内部儿童商业规划与设计研究——以南京部分购物中心为例	张 雷
于晓彤	当代建筑师的中国乡土营建实践研究	张 雷
尤逸尘	南京江宁旭日龙山精品民宿项目（地块2）	张 雷
徐天驹	松阳陈家铺村乡土艺术酒店——1号楼农宅改造设计	张 雷
梁万富	南京江宁旭日龙山精品民宿项目（地块1）	张 雷
陆扬帆	信息时代公共图书馆泛阅览空间设计研究	冯金龙
徐 麟	创意产业园建筑空间模糊性设计研究	冯金龙
杨玉菡	涟水县图书馆建筑设计	冯金龙
王 琳	墨尔本唐人街的中国临终关怀医院设计	冯金龙
林 治	公寓建筑的装配式设计研究——以兴智科技园人才公寓为例	冯金龙
顾一蝶	中北学院学术交流中心建筑设计	冯金龙
车俊颖	面向性能分析的"减法"形体的参数化建模方法研究	吉国华
梁耀波	基于结构性能的形式生成研究——以壳体结构为例	吉国华
王 政	基于空调负荷的多层办公建筑平面优化设计研究	吉国华
张 楠	基于天然采光的生成优化设计技术与方法研究	吉国华
陈 曦	既有工业建筑保护性再利用中围护结构节能策略研究	傅 筱
李天骄	幼儿园建筑生活单元天然采光优化设计研究——以十二班幼儿园为例	傅 筱
岳海旭	台州文化创意产业集聚区3号地块人才公寓区规划与单体改造设计	傅 筱
王冰卿	南京小铜银巷地块超高层建筑前期设计研究	傅 筱
刘 宇	浙江省台州市文化创意产业集聚区旧厂房改造设计——1号地块电子商务园	傅 筱
武春洋	浙江省台州市创意产业集聚区旧厂房改造设计——2号地块汽车研发办公区	傅 筱
谢锡淡	乡村建设模式探究——以杨桥村保护更新为例	周 凌
姚 梦	南京门西凤游寺地块历史信息与空间复原研究	周 凌

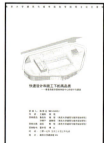

研究生姓名	研究生论文标题	导师姓名
刘 晨	高密度非均质肌理切片的形态指标和天空开阔度的相关性研究	丁沃沃
孙雅贤	城市肌理的平均天空开阔度计算参数设置研究	丁沃沃
郑 伟	公建地块建筑布局规律研究——以南京为例	丁沃沃
吴昇奕	基于市民出行便捷的南京主城区常规公交系统评估	丁沃沃
徐思恒	城市立交桥及其周边城市空间形态研究——以南京为例	华晓宁
林伟圳	关于大学校园空间高差问题与优化策略研究——以南京大学鼓楼校区为例	赵 辰
刘思彤	中国近现代居住类历史街区的保护与再利用研究——以南京为例	赵 辰
吴书其	闽东北传统建造体系现代化更新之卫浴单元设计研究——以福建屏南北村为例	赵 辰
夏候蓉	基于现代交通重整的山地聚落水尾空间景观修复设计——以岭腰村水尾节点设计为例	赵 辰
陈晓敏	传统村落复兴中的村口设计研究——以福建政和岭腰锦屏村村口设计为例	赵 辰
陈修远	快速设计和施工下的高品质——歌华营地体验中心的设计与建造	王骏阳
杨 悦	南京长江大桥建筑形象的创作与接受研究	鲁安东
陈博宇	基于遗传算法的建筑生成设计——以冰橇赛道遮阳设计为例	鲁安东
王曙光	南京市务本蚕种场研究及其更新改造	鲁安东
姚晨阳	城市局部气候分区划分的网格尺寸的合理性研究——以南京市主城区为例	童滋雨
张 进	基于互承结构的厂房改造更新设计研究——以泰州高港某厂房建筑为例	童滋雨
林 陈	文艺复兴时期建筑模型的运用——从伯鲁乃涅斯基到帕拉第奥	胡 恒
王 健	约翰·海杜克的城市理念初探	胡 恒
单泓景	养老建筑天然光非视觉效应研究——以南京市老年公寓为例	吴 蔚
查新彧	太阳能烟囱的通风效果与节能效果研究	秦孟昊
张明杰	热湿缓冲现象及其对建筑能耗影响的研究	秦孟昊

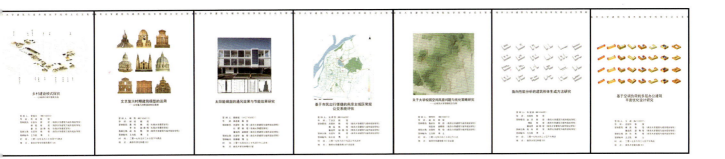

在校学生名单
List of Students

本科生 Undergraduate

2013级学生 / Students 2013

曹舒琪 CAO Shuqi	黄婉莹 HUANG Wanying	罗晓东 LUO Xiaodong	王青 WANG Qing	徐家炜 XU Jiawei	周怡 ZHOU Yi
陈露 CHEN Lu	黄追日 HUANG Zhuiri	吕童 LV Tong	王秋锐 WANG Qiurui	徐瑜灵 XU Yuling	
董素宏 DONG Suhong	吉雨心 JI Yuxin	楠田康雄 KUSUDA YASUO	王瑶 WANG Yao	杨蕾 YANG Lei	
郭金未 GUO Jinwei	贾奕超 JIA Yichao	宋宇珥 SONG Yuxun	王智伟 WANG Zhiwei	章太雷 ZHANG Tailei	
郭硕 GUO Shuo	林之音 LIN Zhiyin	谭皓 TAN Hao	武波 WU Bo	赵焦 ZHAO Jiao	
贺唯嘉 HE Weijia	刘稷祺 LIU Jiqi	涂成祥 TU Chengxiang	夏凡琦 XIA Fanqi	赵梦娣 ZHAO Mengdi	
胡慧慧 HU Huihui	鲁晴 LU Qing	王成阳 WANG Chengyang	夏楠 XIA Nan	赵中石 ZHAO Zhongshi	

2014级学生 / Students 2014

蔡英杰 CAI Yingjie	林宇 LIN Yu	施孝萱 SHI Xiaoxuan	谢峰 XIE Feng	
曹焱 CAO Yan	刘畅 LIU Chang	宋怡 SONG Yi	严紫微 Yan Ziwei	
陈妍霓 CHEN Yanni	刘宛莹 LIU Wanying	宋宇宁 SONG Yuning	杨云睿 YANG Yunrui	
杜孟泽杉 DU Mengzeshan	刘为尚 LIU Weishang	宋云龙 SONG Yunlong	杨钊 YANG Zhao	
胡皓捷 HU Haojie	卢鼎 LU Ding	唐萌 TANG Meng	尹子晗 YIN Zihan	
兰阳 LAN Yang	马西伯 MA Xibo	完颜尚文 WANYAN Shangwen	张俊 ZHANG Jun	
梁晓蕊 LIANG Xiaorui	施少鋆 SHI Shaojun	夏心雨 XIA Xinyu	张珊珊 ZHANG Shanshan	

2015级学生 / Students 2015

卞秋怡 BIAN Qiuyi	陈景杨 Chen Jingyang	李心仪 LI Xinyi	沈静雯 SHEN Jingwen	杨鑫毓 YANG Xinyu
吕文倩 LV Wenqian	邸晓宇 DI Xiaoyu	刘博 LIU Bo	汪榕 WANG Rong	杨洋 YANG Yang
龚正 GONG Zheng	丁展图 DING Zhantu	刘秀秀 LIU Xiuxiu	王晨 WANG Chen	叶庆锋 YE Qingfeng
顾卓琳 GU Zhuolin	顾梦婕 GU Mengjie	刘越 LIU Yue	王雪梅 WANG Xuemei	张昊阳 ZHANG Haoyang
戴添趣 DAI Tianqu	何璇 HE Xuan	罗逍遥 LUO Xiaoyao	卫斌 WEI Bin	周杰 ZHOU Jie
赵彤 ZHAO Tong	兰贤元 LAN Xianyuan	毛志敏 MAO Zhimin	仙海斌 XIAN Haibin	
罗紫璇 LUO Zixuan	李博文 LI Bowen	秦伟航 QIN Weihang	徐玲丽 XU Lingli	

2016级学生 / Students 2016

陈帆 CHEN Fan	黄靖绮 HUANG Jingqi	丘雨辰 QIU Yuchen	于文爽 YU Wenshuang	
陈鸿帆 CHEN Hongfan	黄文凯 HUANG Wenkai	石雪松 SHI Xuesong	余沁蔓 YU Qinman	
陈靖秋 CHEN Jingqiu	雷畅 LEI Chang	司昌尧 SI Changyao	张涵筱 ZHANG Hanxiao	
陈铭行 CHEN Mingxing	李宏健 LI Hongjian	王路 WANG Lu	周子琳 ZHOU Zilin	
陈予婧 CHEN Yujing	李舟涵 LI Zhouhan	吴林天池 WU Lintianchi	朱凌云 ZHU Lingyun	
封翘 FENG Qiao	马彩霞 MA Caixia	吴敏婷 WU Minting		
龚之璇 GONG Zhixuan	潘博 PAN Bo	谢琳娜 XIE Linna		

研究生 Postgraduate

奥珅颖 AO Shenying	雷冬雪 LEI Dongxue	孟文儒 MENG Wenru	王珊珊 WANG Shanshan	许伯晗 XU Bohan	郭瑛 GUO Ying	力振球 LI Zhenqiu	孙昕 SUN Xin	徐蕾 XU Lei
陈观兴 CHEN Guanxing	李彤 LI Tong	潘柳青 PAN Liuqing	魏江洋 WEI Jiangyang	许骏 XU Jun	郭耘锦 GUO Yunjin	刘莹 LIU Ying	谭发兵 TAN Fabing	徐沁心 XU Qinxin
陈相营 CHEN Xiangying	李招成 LI Zhaocheng	潘幼健 PAN Youjian	吴超楠 WU Chaonan	曹峥 CAO Zheng	季平 JI Ping	刘玉婧 LIU Yujing	汤建华 TANG Jianhua	徐婉迪 XU Wandi
段艳文 DUAN Yanwen	刘佳 LIU Jia	沙吉敏 SHA Jimin	吴嘉鑫 WU Jiaxin	陈逸 CHEN Yi	贾江南 JIA Jiangnan	柳筱娴 LIU Xiaoxian	王斌鹏 WANG Binpeng	张楠 ZHANG Nan
符靓璇 FU Jingxuan	刘彦辰 LIU Yanchen	谭子龙 TAN Zilong	夏炎 XIA Yan	仇高颖 QIU Gaoying	姜智 JIANG Zhi	沈康惠 SHEN Kanghui	王晗 WANG Han	赵阳 ZHAO Yang
黄龙辉 Huang Longhui	吕航 LV Hang	王淡秋 WANG Danqiu	肖霄 XIAO Xiao	戴波 DAI Bo	蒯冰清 KUAI Bingqing	施伟 SHI Wei	王倩 WANG Qian	周荣楼 ZHOU Ronglou
姜伟杰 JIANG Weijie	毛军列 MAO Junlie	王冬雪 WANG Dongxue	徐少敏 XU Shaomin	费日晓 FEI Rixiao	李昭 LI Zhao	孙冠成 SUN Guancheng	吴宾 WU Bin	

车俊颖 CHE Junying	顾一蝶 GU Yidie	梁万富 LIANG Wanfu	刘文沛 LIU Wenpei	孙雅贤 SUN Yaxian	王政 WANG Zheng	徐麟 XU Lin	杨玉菡 YANG Yuhan	张丛 ZHANG Cong
陈博宇 CHEN Boyu	韩书园 HAN Shuyuan	梁耀波 LIANG Yaobo	刘宇 LIU Yu	谭健 TAN Jian	武春洋 WU Chunyang	徐思恒 XU Siheng	姚晨阳 YAO Chenyang	张海宁 ZHANG Haining
陈凌杰 CHEN Lingjie	胡任元 HU Renyuan	林陈 LIN Chen	陆扬帆 LU Yangfan	田金华 TIAN Jinhua	吴昇奕 WU Shengyi	徐天驹 XU Tianju	姚梦 YAO Meng	张明杰 ZHANG Mingjei
陈曦 CHEN Xi	黄广伟 HUANG Guangwei	林伟圳 LIN Weizhen	骆国建 LUO Guojian	王冰卿 WANG Bingqing	吴书其 WU Shuqi	许文韬 XU Wentao	尤逸尘 YOU Yichen	张进 ZHANG Jin
陈晓敏 CHEN Xiaomin	蒋西亚 JIANG Xiya	林治 LIN Zhi	宁凯 NING Kai	王健 WANG Jian	吴婷婷 WU Tingting	徐晏 XU Yan	于晓彤 YU Xiaotong	张楠 ZHANG Nan
陈修远 CHEN Xiuyuan	焦宏建 JIAO Hongbin	刘晨 LIU Chen	彭蕊寒 PENG Ruihan	王琳 WANG Lin	夏候蓉 XIA Hourong	杨天仪 YANG Tianyi	岳海旭 YUE Haixu	张强 ZHANG Qiang
程斌 CHENG Bin	李天骄 LI Tianjiao	刘芮 LIU Rui	单泓景 SHAN Hongjing	王曙光 WANG Shuguang	谢锡淡 XIE Xidan	杨悦 YANG Yue	查新彧 ZHA Xinyu	郑伟 ZHENG Wei
高翔 GAO Xiang	廉英豪 LIAN Yinghao	刘思彤 LIU Sitong						

曹阳 CAO Yang	胡珊 HU Shan	彭丹丹 PENG Dandan	吴松霖 WU Songlin	赵婧靓 ZHAO Jingliang	蒋建昕 JIANG Jianxin	施成 SHI Cheng	于慧颖 YU Huiying	赵伟 ZHAO Wei
陈祺 CHEN Qi	黄丽 HUANG Li	邵思宇 SHAO Siyu	席弘 XI Hong	周剑晖 ZHOU Jianhui	蒋造时 JIANG Zaoshi	宋春亚 SONG Chunya	余星凯 XU Xingkai	种桂梅 ZHONG Guimei
程思远 CHENG Siyuan	贾福龙 JIA Fulong	沈佳磊 SHEN Jialei	徐一品 XU Yipin	艾心 AI Xin	黎乐源 LI Leyuan	宋富敏 SONG Fumin	张本纪 ZHANG Benji	周明辉 ZHOU Minghui
迟海韵 CHI Haiyun	蒋佳瑶 JIANG Jiayao	拓展 TUO Zhan	徐亦杨 XU Yiyang	陈嘉铮 CHEN Jiazheng	刘泽超 LIU Zechao	王晓茜 WANG Xiaoqian	张豪杰 ZHANG Haojie	周贤春 ZHOU Xianchun
崔傲寒 CUI Aohan	蒋靖才 JIANG Jingcai	王敏姣 WANG Minjiao	杨益晖 YANG Yihui	陈立华 CHEN Lihua	柳纬宇 LIU Weiyu	吴结松 WU Jiesong	张靖 ZHANG Jing	周洋 ZHOU Yang
方飞 FANG Fei	李若尧 LI Ruoyao	王却奁 WANG Quelian	杨肇伦 YANG Zhaolun	黄凯峰 HUANG Kaifeng	吕秉田 LV Bingtian	谢星宇 XIE Xingyu	张黎萌 ZHANG Limeng	邹晓蕾 ZOU Xiaolei
冯琪 FENG Qi	李文凯 LI Wenkai	王峥涛 WANG Zhengtao	张洪光 ZHANG Hongguang	江振彦 JIANG Zhenyan	缪姣姣 MIAO Jiaojiao	谢忠雄 XIE Zhongxiong	张学 ZHANG Xue	刘晓君 LIU Xiaojun
顾聿笙 GU Yusheng	刘茏鑫 LIU Longxin	王子珊 WANG Zishan	张欣 ZHANG Xin	姜澜 JIANG Lan	沈珊珊 SHEN Shanshan			

陈硕 CHEN Shuo	董晶晶 DONG Jingjing	黄子恩 HUANG Zi'en	梁庆华 LIANG Qinghua	聂柏慧 NIE Baihui	王浩哲 WANG Haozhe	文涵 WEN Han	徐新杉 XU Xinshan	臧倩 ZANG Qian
曹永青 CAO Yongqing	高祥震 GAO Xiangzhen	季惠敏 JI Huimin	刘刚 LIU Gang	戚迹 QI Ji	王丽丽 WANG Lili	吴帆 WU Fan	徐雅甜 XU Yatian	张馨元 ZHANG Xinyuan
陈思涵 CHEN Sihan	葛嘉许 GE Jiaxu	蒋玉若 JIANG Yuruo	刘江全 LIU Jiangquan	裘嘉珺 QIU Jiajun	王姝宁 WANG Shuning	吴家禾 WU Jiahe	杨瑞东 YANG Ruidong	章程 ZHANG Cheng
陈欣冉 CHEN Xinran	耿蒙蒙 GENG Mengmeng	金璐璐 JIN Lulu	刘宣 LIU Xuan	芮丽燕 RUI Liyan	王坦 WANG Tan	吴桐 WU Tong	杨喆 YANG Zhe	赵霏霏 ZHAO Feifei
陈妍 CHEN Yan	宫传佳 GONG Chuanjia	李鹏程 LI Pengcheng	刘姿佑 LIU Ziyou	苏彤 SU Tong	王婷婷 WANG Tingting	吴峥嵘 WU Zhengrong	于明霞 YU Mingxia	赵媛情 ZHAO Yuanqian
程睿 CHENG Rui	桂瑜 GUI Yu	李恬楚 LI Tianchu	娄弯弯 LOU wanwan	孙鸿腾 SUN Hongpeng	王一侬 WANG Yinong	谢灵晋 XIE Lingjin	于昕 YU Xin	朱鼎祥 ZHU Dingxiang
从彬 CONG Bin	郭嫦嫦 GUO Changchang	李伟 LI Wei	马亚菲 MA Yafei	汤晋 TANG Jin	王永 WANG Yong	熊攀 XIONG Pan	袁一 YUAN Yi	朱凌峥 ZHU Lingzheng
代晓荣 DAI Xiaorong	黄陈瑶 HUANG Chenyao	李潇乐 LI Xiaole	梅凯强 MEI Kaiqiang	童月清 TONG Yueqing	王照宇 WANG Zhaoyu	徐沙 XU Sha	袁子燕 YUAN Ziyan	

图书在版编目（CIP）数据

南京大学建筑与城市规划学院建筑系教学年鉴. 2016—2017 / 王丹丹编. -- 南京：东南大学出版社，2017.12
ISBN 978-7-5641-7512-2

Ⅰ. ①南… Ⅱ. ①王… Ⅲ. ①建筑学—教学研究—高等学校—南京—2016—2017—年鉴②城市规划—教学研究—高等学校—南京—2016—2017—年鉴 Ⅳ. ①TU-42

中国版本图书馆CIP数据核字（2017）第293222号

编 委 会：	丁沃沃　赵　辰　吉国华　周　凌　王丹丹
装帧设计：	王丹丹　丁沃沃
版面制作：	梁晓蕊　刘宛莹　施孝萱　宋　怡
参与制作：	颜骁程　陶敏悦
责任编辑：	姜　来　魏晓平

出版发行：	东南大学出版社
社　　址：	南京市四牌楼2号
出 版 人：	江建中
网　　址：	http://www.seupress.com
邮　　箱：	press@seupress.com
邮　　编：	210096
经　　销：	全国各地新华书店
印　　刷：	南京新世纪联盟印务有限公司
开　　本：	787mm×1092mm　1/20
印　　张：	9.5
字　　数：	352千
版　　次：	2017年12月第1版
印　　次：	2017年12月第1次印刷
书　　号：	ISBN 978-7-5641-7512-2
定　　价：	68.00元

本社图书若有印装质量问题，请直接与营销部联系。电话：025-83791830